COHESION/ADHE

TASK CARD SERIES

Conceived and written by

RON MARSON

Illustrated by

PEG MARSON

LEARNING SYSTEMS

342 S Plumas Street
Willows, CA 95988

www.topscience.org

WHAT CAN YOU COPY?

Dear Educator,

Please honor our copyright restrictions. We offer liberal options and guidelines below with the intention of balancing your needs with ours. When you buy these labs and use them for your own teaching, you sustain our work. If you "loan" or circulate copies to others without compensating TOPS, you squeeze us financially, and make it harder for our small non-profit to survive. Our well-being rests in your hands. Please help us keep our low-cost, creative lessons available to students everywhere. Thank you!

PURCHASE, ROYALTY and LICENSE OPTIONS

TEACHERS, HOMESCHOOLERS, LIBRARIES:

We do all we can to keep our prices low. Like any business, we have ongoing expenses to meet. We trust our users to observe the terms of our copyright restrictions. While we prefer that all users purchase their own TOPS labs, we accept that real-life situations sometimes call for flexibility.

Reselling, trading, or loaning our materials is prohibited unless one or both parties contribute an Honor System Royalty as fair compensation for value received. We suggest the following amounts – let your conscience be your guide.

HONOR SYSTEM ROYALTIES: If making copies from a library, or sharing copies with colleagues, please calculate their value at 50 cents per lesson, or 25 cents for homeschoolers. This contribution may be made at our website or by mail (addresses at the bottom of this page). Any additional tax-deductible contributions to make our ongoing work possible will be accepted gratefully and used well.

Please follow through promptly on your good intentions. Stay legal, and do the right thing.

SCHOOLS, DISTRICTS, and HOMESCHOOL CO-OPS:

PURCHASE Option: Order a book in quantities equal to the number of target classrooms or homes, and receive quantity discounts. If you order 5 books or downloads, for example, then you have unrestricted use of this curriculum for any 5 classrooms or families per year for the life of your institution or co-op.

2-9 copies of any title: 90% of current catalog price + shipping.

10+ copies of any title: 80% of current catalog price + shipping.

ROYALTY/LICENSE Option: Purchase just one book or download *plus* photocopy or printing rights for a designated number of classrooms or families. If you pay for 5 additional Licenses, for example, then you have purchased reproduction rights for an entire book or download edition for any **6** classrooms or families per year for the life of your institution or co-op.

1-9 Licenses: 70% of current catalog price per designated classroom or home.

10+ Licenses: 60% of current catalog price per designated classroom or home.

WORKSHOPS and TEACHER TRAINING PROGRAMS:

We are grateful to all of you who spread the word about TOPS. Please limit copies to only those lessons you will be using, and collect all copyrighted materials afterward. No take-home copies, please. Copies of copies are strictly prohibited.

For licensing, honor system royalty payments, contact: **www.TOPScience.org**; or **TOPS Learning Systems 342 S Plumas St, Willows CA 95988**; or inquire at **customerservice@topscience.org**

ISBN 978-0-941008-83-9

CONTENTS

INTRODUCTION

TEACHING NOTES

REPRODUCIBLE MATERIALS

A TOPS Model for Effective Science Teaching...

If science were only a set of explanations and a collection of facts, you could teach it with blackboard and chalk. You could assign students to read chapters and answer the questions that followed. Good students would take notes, read the text, turn in assignments, then give you all this information back again on a final exam. Science is traditionally taught in this manner. Everybody learns the same body of information at the same time. Class togetherness is preserved.

But science is more than this.

Science is also process — a dynamic interaction of rational inquiry and creative play. Scientists probe, poke, handle, observe, question, think up theories, test ideas, jump to conclusions, make mistakes, revise, synthesize, communicate, disagree and discover. Students can understand science as process only if they are free to think and act like scientists, in a classroom that recognizes and honors individual differences.

Science is *both* a traditional body of knowledge *and* an individualized process of creative inquiry. Science as process cannot ignore tradition. We stand on the shoulders of those who have gone before. If each generation reinvents the wheel, there is no time to discover the stars. Nor can traditional science continue to evolve and redefine itself without process. Science without this cutting edge of discovery is a static, dead thing.

Here is a teaching model that combines the best of both elements into one integrated whole. It is only a model. Like any scientific theory, it must give way over time to new and better ideas. We challenge you to incorporate this TOPS model into your own teaching practice. Change it and make it better so it works for you.

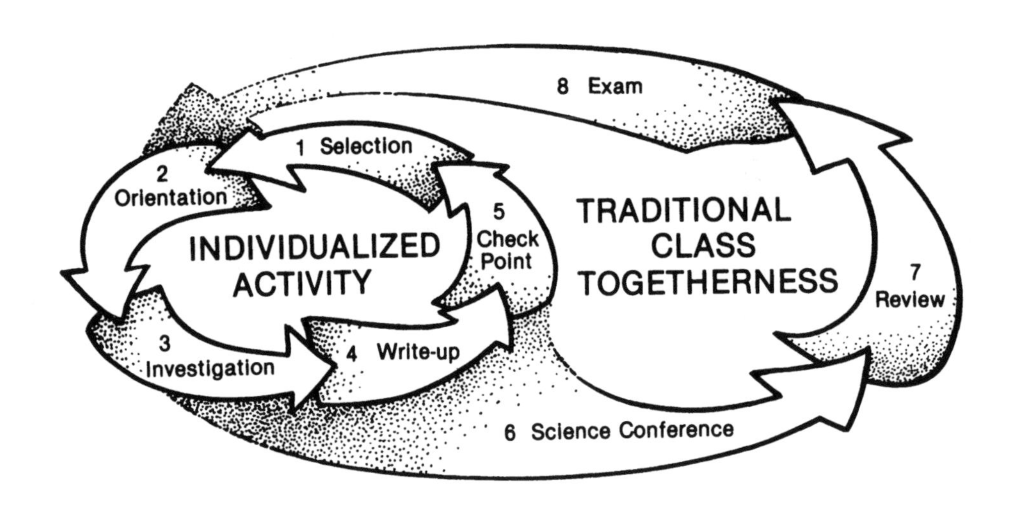

1. SELECTION

Doing TOPS is as easy as selecting the first task card and doing what it says, then the second, then the third, and so on. Working at their own pace, students fall into a natural routine that creates stability and order. They still have questions and problems, to be sure, but students know where they are and where they need to go.

Students generally select task cards in sequence because new concepts build on old ones in a specific order. There are, however, exceptions to this rule: students might *skip* a task that is not challenging; *repeat* a task with doubtful results; *add* a task of their own design to answer original "what would happen if" questions.

2. ORIENTATION

Many students will simply read a task card and immediately understand what to do. Others will require further verbal interpretation. Identify poor readers in your class. When they ask, "What does this mean?" they may be asking in reality, "Will you please read this card aloud?"

With such a diverse range of talent among students, how can you individualize activity and still hope to finish this module as a cohesive group? It's easy. By the time your most advanced students have completed all the task cards, including the enrichment series at the end, your slower students have at least completed the basic core curriculum. This core provides the common background so necessary for meaningful discussion, review and testing on a class basis.

3. INVESTIGATION

Students work through the task cards independently and cooperatively. They follow their own experimental strategies and help each other. You encourage this behavior by helping students only *after* they have tried to help themselves. As a resource person, you work to stay *out* of the center of attention, answering student questions rather than posing teacher questions.

When you need to speak to everyone at once, it is appropriate to interrupt individual task card activity and address the whole class, rather than repeat yourself over and over again. If you plan ahead, you'll find that most interruptions can fit into brief introductory remarks at the beginning of each new period.

4. WRITE-UP

Task cards ask students to explain the "how and why" of things. Write-ups are brief and to the point. Students may accelerate their pace through the task cards by writing these reports out of class.

Students may work alone or in cooperative lab groups. But each one must prepare an original write-up. These must be brought to the teacher for approval as soon as they are completed. Avoid dealing with too many write-ups near the end of the module, by enforcing this simple rule: each write-up must be approved *before* continuing on to the next task card.

5. CHECK POINT

The student and teacher evaluate each write-up together on a pass/no-pass basis. (Thus no time is wasted haggling over grades.) If the student has made reasonable effort consistent with individual ability, the write-up is checked off on a progress chart and included in the student's personal assignment folder or notebook kept on file in class.

Because the student is present when you evaluate, feedback is immediate and effective. A few seconds of this direct student-teacher interaction is surely more effective than 5 minutes worth of margin notes that students may or may not heed. Remember, you don't have to point out every error. Zero in on particulars. If reasonable effort has not been made, direct students to make specific improvements, and see you again for a follow-up check point.

A responsible lab assistant can double the amount of individual attention each student receives. If he or she is mature and respected by your students, have the assistant check the even-numbered write-ups while you check the odd ones. This will balance the work load and insure that all students receive equal treatment.

6. SCIENCE CONFERENCE

After individualized task card activity has ended, this is a time for students to come together, to discuss experimental results, to debate and draw conclusions. Slower students learn about the enrichment activities of faster students. Those who did original investigations, or made unusual discoveries, share this information with their peers, just like scientists at a real conference. This conference is open to films, newspaper articles and community speakers. It is a perfect time to consider the technological and social implications of the topic you are studying.

7. READ AND REVIEW

Does your school have an adopted science textbook? Do parts of your science syllabus still need to be covered? Now is the time to integrate other traditional science resources into your overall program. Your students already share a common background of hands-on lab work. With this shared base of experience, they can now read the text with greater understanding, think and problem-solve more successfully, communicate more effectively.

You might spend just a day on this step or an entire week. Finish with a review of key concepts in preparation for the final exam. Test questions in this module provide an excellent basis for discussion and study.

8. EXAM

Use any combination of the review/test questions, plus questions of your own, to determine how well students have mastered the concepts they've been learning. Those who finish your exam early might begin work on the first activity in the next new TOPS module.

Now that your class has completed a major TOPS learning cycle, it's time to start fresh with a brand new topic. Those who messed up and got behind don't need to stay there. Everyone begins the new topic on an equal footing. This frequent change of pace encourages your students to work hard, to enjoy what they learn, and thereby grow in scientific literacy.

Getting Ready

Here is a checklist of things to think about and preparations to make before your first lesson.

✔ Review the scope and sequence.

Take just a few minutes, right now, to page through each activity. Read each *Task Objective* at the top of the page and scan each task card underneath.

✔ Set aside appropriate class time.

Allow an average of perhaps 1 class period per lesson (more for younger students), plus time at the end of this module for discussion, review and testing. If your schedule doesn't allow this much science, consult the logic tree on page E to see which activities you can safely omit without breaking conceptual links between lessons.

✔ Number your task card masters.

The small number printed in the lower right corner of each task card shows its position within the series. If this ordering fits your schedule, copy each number into the blank parentheses directly above it at the top of the card. Be sure to use pencil; you may decide to revise, rearrange, add or omit task cards the next time you teach this module. Simply insert your own better ideas on 4 x 6 index cards wherever they fit best, and renumber your sequence. This allows your curriculum to adapt and grow as you do.

✔ Photocopy task cards for student use.

Decide how you want to distribute task cards among your students. Find a list of management options opposite the task card master numbered "cards 1-2." Then photocopy and collate classroom sets of task cards in the numbers you need, plus supplementary pages (if any) at the back of this module. The *Materials* list that accompanies each activity specifies when these supplementary pages are required, and how many to photocopy.

We allow you to photocopy all permissible materials, as long as you limit the distribution of copies you make to the students you personally teach. Encourage other teachers who want to use this module to purchase their own TOPS module. This supports TOPS financially, enabling us to continue publishing new modules for you.

✔ Collect needed materials.

Please see page D, opposite, for details.

✔ Organize a way to track assignments.

Keep student work on file in class. If you lack a file cabinet, a box with a brick will serve. File folders or notebooks both make suitable assignment organizers. Students will feel a sense of accomplishment as they see their folders grow heavy, or their notebooks fill, with completed assignments. Since all papers stay together, reference and review are facilitated.

Ask students to number a sheet of paper 1, 2, 3..., to the total number of task cards in this module, then tape it inside the front cover of their folders or notebooks. Track individual progress through this module (and future modules) by initialing lesson numbers as students complete each assignment.

✔ Review safety procedures.

In our litigation-conscious society, we find that publishers are often more committed to protecting themselves from liability suits than protecting students from physical hazards. Lab instructions are too often filled with spurious advisories, cautions and warnings that desensitize students to safety in general. If we cry "Wolf!" too often, real warnings of present danger may go unheeded.

At TOPS we endeavor to use good sense in deciding what students already know (don't stab yourself in the eye) and what they should be told (don't look directly at the sun.) Scissors and pins, of course, could be dangerous in the hands of unsupervised children. Nor can this curriculum anticipate irresponsible behavior or negligence. As the teacher, it is ultimately your responsibility to see that common-sense safety rules are followed; your students' responsibility is to respect and protect themselves and each other.

Begin with these basic safety rules:

- Eye Protection: Wear safety goggles when heating liquids or solids to high temperatures.
- Poisons: Never taste anyting unless you are told to do so.
- Fire: Keep loose hair or clothing away from flames. Point test tubes which are heating away from your face and your neighbor's.
- Glass Tubing: Don't force through stoppers. (The teacher should fit glass tubes to stoppers in advance, using a lubricant.)
- Gas: Light the match first, before turning on the gas.

✔ Communicate your grading expectations.

Whatever your grading philosophy, your students need to understand how they will be assessed. Here is a scheme that counts individual effort, attitude and overall achievement. We think these 3 components deserve equal weight:

- Pace (effort): Tally the number of check points and extra credit experiments you have initialed for each student. Low-ability students should be able to keep pace with gifted students, since write-ups are evaluated relative to individual performance standards on a pass/no-pass basis. Students with absences or those who tend to work slowly might assign themselves more homework out of class.
- Participation (attitude): This is a subjective grade, assigned to measure personal initiative and responsibility. Active participators who work to capacity receive high marks. Inactive onlookers who waste time in class and copy the results of others receive low marks.
- Exam (achievement): Activities point toward generalizations that provide a basis for hypothesizing and predicting. Review/Test questions beginning on page G will help you assess whether students understand relevant theory and can apply it in a predictive way.

Gathering Materials

Listed below is everything you'll need to teach this module. You already have many of these items. The rest are available from your supermarket, drugstore and hardware store. Laboratory supplies may be ordered through a science supply catalog.

Keep this classification key in mind as you review what's needed:

special in-a-box materials:	general on-the-shelf materials:
Italic type suggests that these materials are unusual. Keep these specialty items in a separate box. After you finish teaching this module, label the box for storage and put it away, ready to use again the next time you teach this module.	Normal type suggests that these materials are common. Keep these basics on shelves or in drawers that are readily accessible to your students. The next TOPS module you teach will likely utilize many of these same materials.
(substituted materials):	*optional materials:
Parentheses enclosing any item suggests a ready substitute. These alternatives may work just as well as the original, perhaps better. Don't be afraid to improvise, to make do with what you have.	An asterisk sets these items apart. They are nice to have, but you can easily live without them. They are probably not worth an extra trip to the store, unless you are gathering other materials as well.

Everything is listed in order of first use. Start gathering at the top of this list and work down. Ask students to bring recycled items from home. The teaching notes may occasionally suggest additional student activity under the heading "Extensions." Materials for these optional experiments are listed neither here nor in the teaching notes. Read the extension itself to find out what new materials, if any, are required.

Needed quantities depend on how many students you have, how you organize them into activity groups, and how you teach. Decide which of these 3 estimates best applies to you, then adjust quantities up or down as necessary:

$Q_1/Q_2/Q_3$

- **Single Student:** Enough for 1 student to do all the experiments.
- **Individualized Approach:** Enough for 30 students informally working in pairs, all self-paced.
- **Traditional Approach:** Enough for 30 students, organized into pairs, all doing the same lesson.

KEY: *special in-a-box materials* (substituted materials) general on-the-shelf materials *optional materials

Qty	Item
4/40/40	dropper bottles & droppers
1/3/3	rolls masking tape
1/1/1	bottle blue food coloring with dropper dispenser
1/1/1	bottle 70% rubbing alcohol
1/1/1	source of tap water
1/1/1	*bottle Joy® or Dawn® liquid detergent – see notes 20*
1/1/1	bottle corn oil
1/1/1	roll waxed paper
1/10/10	small graduated cylinders, 10 mL capacity
1/10/10	*hand calculators
1/10/10	pennies
1/1/1	roll soft paper towels
1/10/10	scissors
1/5/10	large jars or equivalent supports – see activity 3
1/1/1	clock with second hand (wristwatches)
1/10/10	paper plates, 9 inch diameter
1/4/10	drinking glasses
1/4/10	shallow bowls
1/1/1	shaker of fine pepper
1/50/50	aluminum straight pins – see notes 7
1/1/1	bar of soap
1/1/1	*pkg solid camphor – see notes 6*
2/20/20	medium styrofoam cups, – see notes 7, 8, 20-23
1/10/10	toothpicks
1/30/30	plastic drinking straws, about ¼ inch diameter
1/10/10	string pieces, at least 20 cm (10 inches) long
4/20/40	microscope slides
1/4/10	candles
1/4/10	paper clips
5/30/50	medium or large baby food jars
4/20/40	thin rubber bands
1/4/10	eyedroppers – notes 11
1/4/10	hand lenses
1/10/10	sheets newspaper – see notes 13
1/3/10	*sets of 8 washable colored markers*
1/3/10	rolls clear tape
1/1/1	shaker of table salt
1/10/10	*plastic tubs with lids – see notes 20*
1/1/1	plastic gallon (4 L) milk jug
1/1/1	*bottle glycerine*
1/1/1	*gallon distilled or deionized water – see notes 20
1/1/1	spool thread
2/6/20	size-D batteries, dead or alive
1/10/10	pieces corrugated cardboard, 5 x 20 cm or larger
1/3/10	index cards, 4 x 6 inch
1/1/1	roll aluminum foil
1/2/10	*medium test tubes, about 1.5 cm diameter
1/2/10	meter sticks
1/1/1	*bottle of vinegar

Sequencing Task Cards

This logic tree shows how all the task cards in this module tie together. In general, students begin at the bottom of the tree and work up through the related branches. As the diagram suggests, upper level activities build on lower level activities.

At the teacher's discretion, certain activities can be omitted, or sequences changed, to meet specific class needs. The only activities that must be completed in sequence are indicated by leaves that open *vertically* into the ones above them. In these cases the lower activity is a prerequisite to the upper.

When possible, students should complete the task cards in the same sequence as numbered. If time is short, however, or certain students need to catch up, you can use the logic tree to identify concept-related *horizontal* activities. Some of these might be omitted, since they serve only to reinforce learned concepts, rather than introduce new ones.

On the other hand, if students complete all the activities at a certain horizontal concept level, then experience difficulty at the next higher level, you might move back down the logic tree to have students repeat specific key activities for greater reinforcement.

For whatever reason, when you wish to make sequence changes, you'll find this logic tree a valuable reference. Parentheses in the upper right corner of each task card allow you total flexibility; they are left blank so you can pencil in sequence numbers of your own choosing.

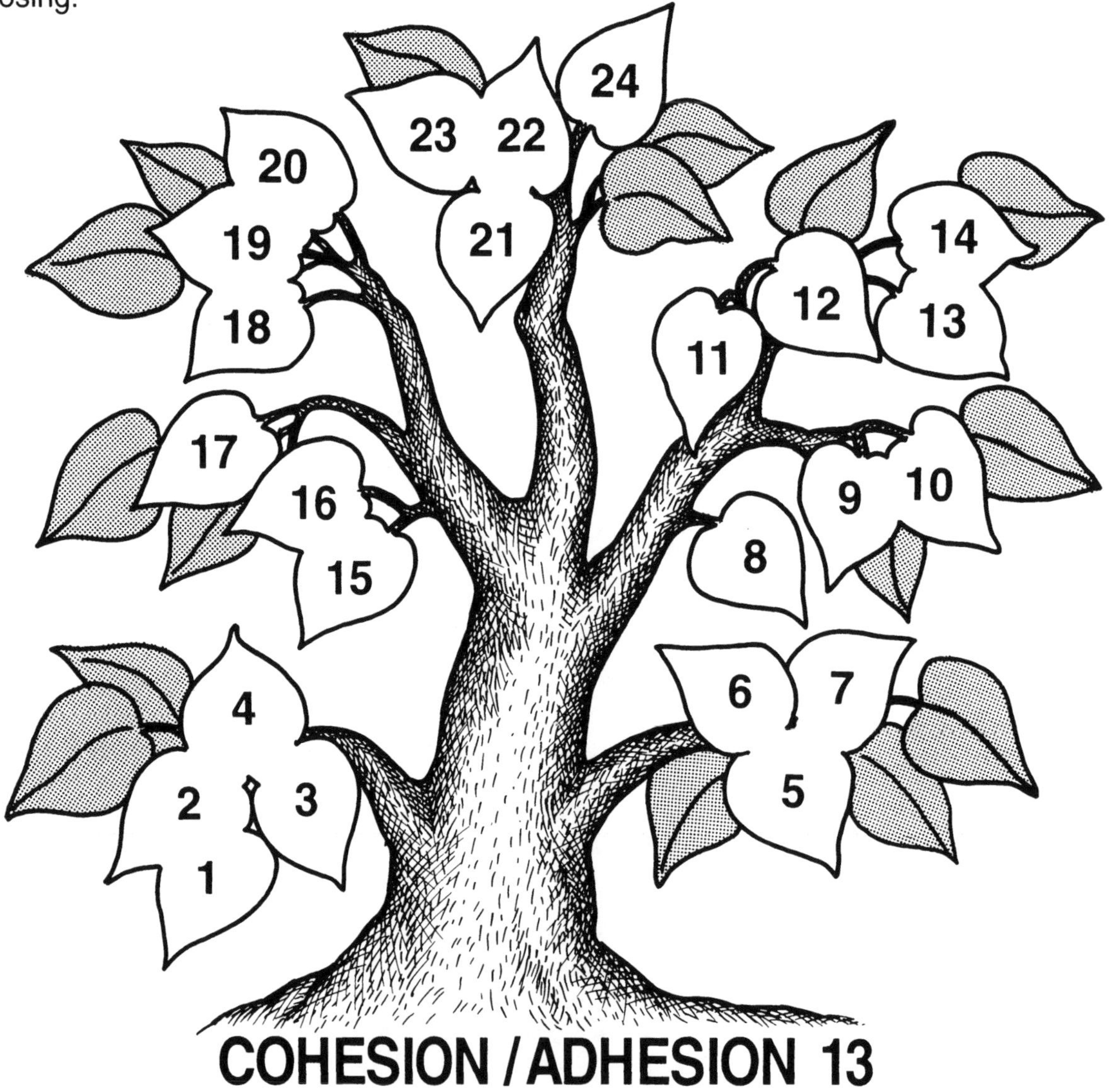

LONG-RANGE OBJECTIVES

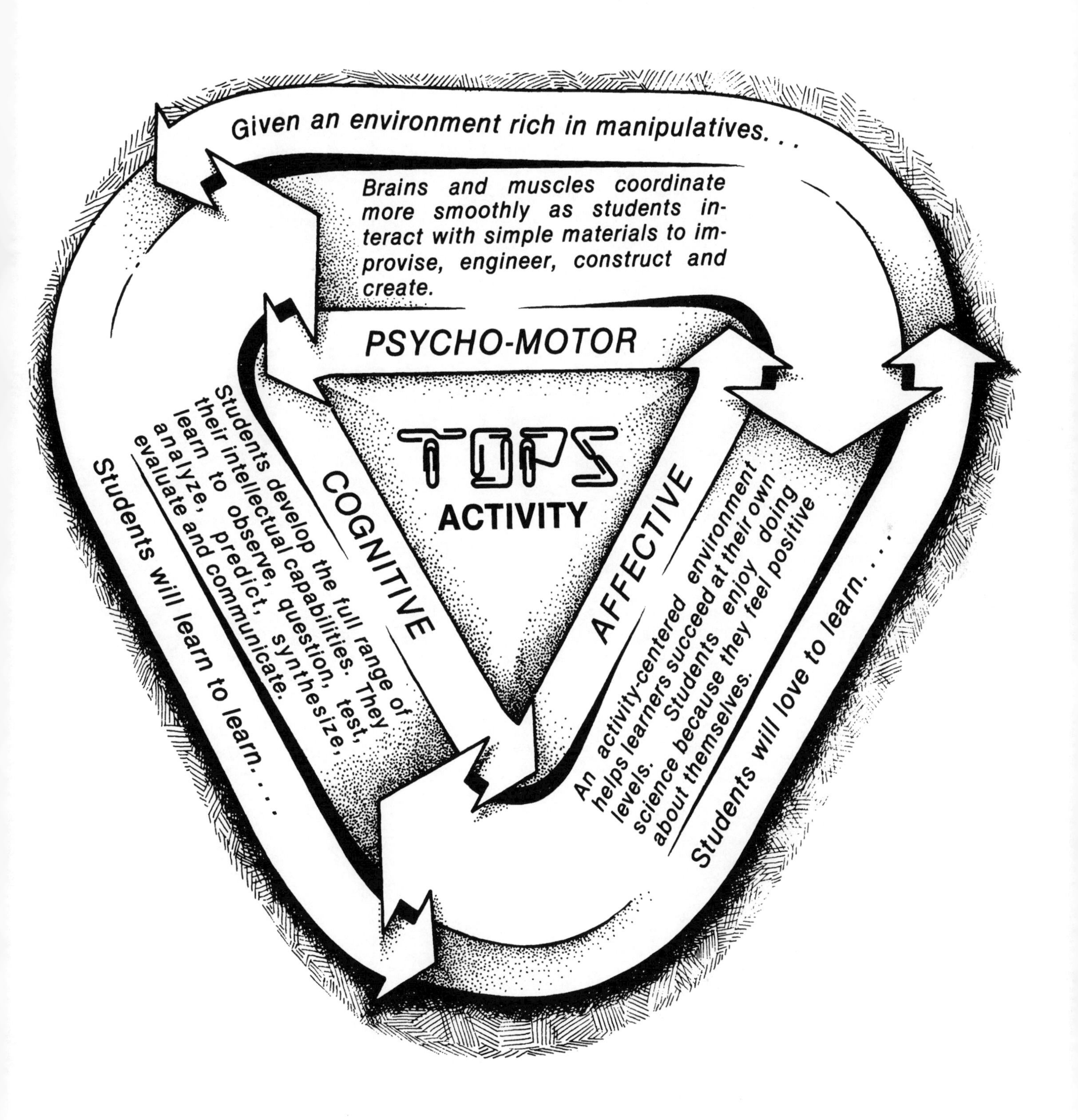

Review / Test Questions

Photocopy the questions below. On a sheet of blank paper, cut and paste those questions you want to use in your test. Include questions of your own design, as well. Place all these questions on a single page for students to answer on another paper, or leave space for student responses after each question, as you wish. Duplicate a class set and your custom-made test is ready to use. Use leftover questions as a review in preparation for the final exam.

tasks 1-2

Accurately draw the water line in each test tube. Test tube **SW** holds as much Soapy Water as possible. Test tube **TW** holds as much Tap Water as possible.

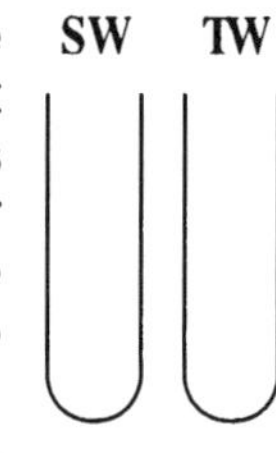

tasks 1-4 A

Where is the cohesion? The adhesion?

a. A drop of rainwater slides down my window.

b. A sand castle holds together as long as the sand is wet.

tasks 1-4 B

Which is stronger, cohesion in the liquid or its adhesion to the solid? Explain.

a. Water on a waxed surface.

b. Oil in a metal engine.

task 5

Account for the strong cohesive forces between water molecules.

tasks 5-6

Predict what happens when you drip a drop of soapy water between 2 toothpicks floating in a bowl of tap water. Give reasons for your prediction.

tasks 5-7

A water strider is a bug that walks on water with dry feet. Explain how hydrogen bonds make this possible.

task 7

A metal paper clip is resting on the surface of a glass of water. If you push it under, will it float back to the surface? Explain.

tasks 8-9

Water tends to dribble down the outside of the spout when poured from a glass pitcher.

a. Why does this happen?

b. How might you use a candle to fix the problem?

tasks 9-10

Rainwater that wets the outside surface of a rock also moistens its interior cracks and crevices. How does this happen?

task 11

Accurately draw the water line in each test tube. Test tube **L** is filled Less than full. Test tube **M** is filled More than full.

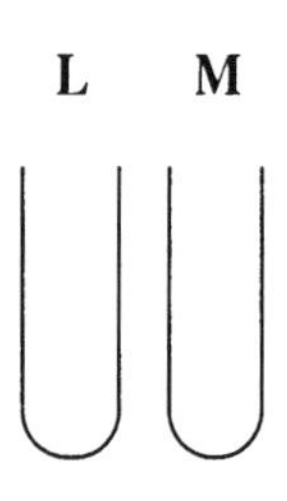

task 12

Explain how a tree moves water from its roots below ground to its leaves high in the air.

tasks 12-13

A paper towel is cut to the same size as a piece of notebook paper. Which soaks up more water? Why?

tasks 13-14

A chromatogram streaks from blue into green into yellow. What color was the original marker? What dye colors does it likely contain?

task 15

Oil is spilled in water. What can you spray on the slick to make it easier to clean up? Explain.

task 16

Two oil slicks reflect light to your eyes, but you see colors in only one of them. How are they different?

task 17

Which is stronger, the surface tension of pure water or the surface tension of soap film? Defend your answer.

task 18

(a)
(b)
(c)

a. Which wave trains are in phase? Out of phase?

b. Which wave trains interfere constructively? Destructively?

tasks 18-20 A

Why does light reflected from a soap bubble produce color?

tasks 18-20 B

You dip the rim of a cup is into bubble solution to create a soapy film, then tilt it at an angle to drain. Soon you see bands of color interference in reflected daylight. Are these bands horizontal or vertical? Why?

tasks 21-22 A

A piece of wire is shaped like a "U" with a piece of thread tied loosely across its middle, then dipped into bubble solution.

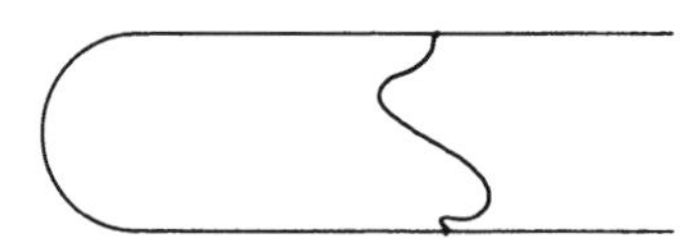

a. Draw the shape of the soap film that forms between wire and thread.

b. Why does the soap film assume this shape and no other?

tasks 21-22 B

You blow a bubble in the air. Why is it spherical?

task 22

Draw side views that show...

a. Two bubbles of equal size united by a common film.

b. A large bubble and small bubble united by a common film.

task 23

Draw top views that show...

a. Two bubbles of equal size united by a common film.

b. One bubble surrounded by three others, all of equal size.

task 24

a. Why do bubble domes break?

b. How can you make them larger?

Answers

tasks 1-2

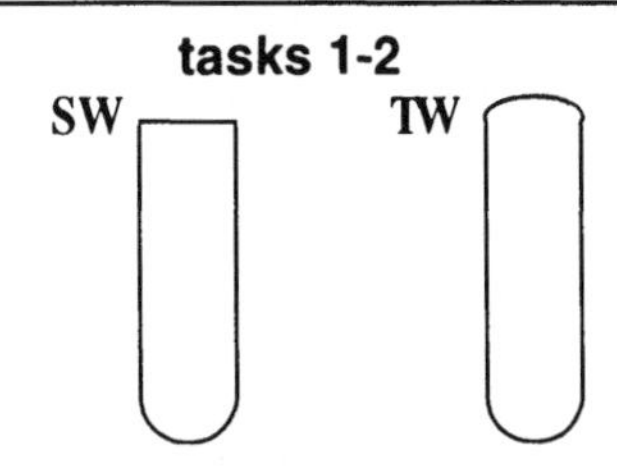

tasks 1-4 A

a. Cohesion holds the water in a single drop as it slides down my window. Adhesion makes the water drop cling to the glass.
b. Adhesion attracts water to sand. Cohesive forces hold water to itself.

tasks 1-4 B

a. The cohesion of water is stronger than its adhesion to a waxed surface, so it to pulls itself into "tall" drops.
b. The cohesion of oil is weaker than its adhesion to metal, coating and protecting all parts of the engine.

task 5

Water molecules are mutually attracted by strong hydrogen bonds. These are caused by electrostatic attraction between electron-rich areas on oxygen atoms and electron-poor areas on hydrogen atoms.

tasks 5-6

The soapy water drop breaks nearby hydrogen bonds on the water's surface. Unbroken hydrogen bonds surrounding this drop pull back on all sides, drawing the toothpicks apart.

tasks 5-7

The water strider stands on the surface "skin" of the water, supported by strong hydrogen bonds that hold the water molecules together. Its feet stay dry because they don't break this network of hydrogen bonds.

task 7

No. The paper clip is resting on the surface "skin" of the water, supported by surface tension. When pushed under water, it breaks through the hydrogen bonds that support it, and it sinks.

tasks 8-9

a. The adhesion of water to glass is strong enough to keep both in continuous contact. Water flows over the lip to the outside of the pitcher rather than pouring freely.
b. Rub the candle on both sides of the glass spout to coat it with wax. Because adhesion between water and wax is low, the water now releases freely from the spout.

tasks 9-10

Water flows from the outside of the rock into its narrow cracks and crevices by capillary action: adhesion attracts water to the dry spaces between rock, while cohesion drags in additional water to resupply the advancing front.

task 11

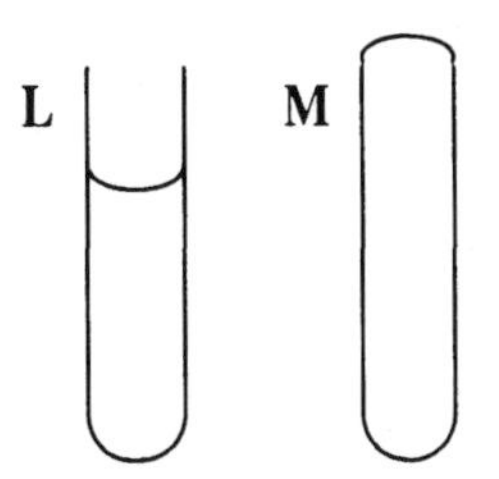

task 12

Ground water flows by capillary action through tiny tubes in the tree roots, upward through capillaries in the trunk and branches, and out through veins in the leaves where it finally evaporates. This establishes a continuous capillary flow up the tree.

tasks 12-13

The paper towel soaks up more water because of its porous texture. Adhesion pulls water into the dry towel on contact, where it advances in tiny cohesive streams through a network of internal capillaries. Notebook paper, by contrast, is smooth and dense with no interior pathways to carry water by capillary action.

tasks 13-14

The original marker was probably colored green. Green in the chromatogram has not yet fully separated into its component primary dyes, colored blue and yellow.

task 15

Spray soap or detergent on the oil slick to reduce the surface tension of water underneath. This allows cohesive forces in the oil to pull it together into more compact drops that are easier to skim away.

task 16

The oil slick that reflects colors is thinner than the one that doesn't.

task 17

The surface tension of pure water must be stronger than the surface tension in soap film. Bubbles neither form nor remain on pure water unless you first weaken its surface tension with the addition of soap.

task 18

a. Wave train (a) is out of phase with wave trains (b) and (c) which are in phase with each other.
b. Wave train (a) interferes destructively with wave trains (b) and (c), which interfere constructively with each other.

tasks 18-20 A

White light, containing all wavelengths in the color spectrum, reflects off both the outer and inner surfaces of a bubble film. Those wave trains that bounce off the inside travel a small distance farther — through the thickness of the bubble film and back again — before rejoining the wave trains that bounce off the outside. This phase-shifts the reflected light that enters your eyes. You see colors as different wavelengths interfere constructively and destructively.

tasks 18-20 B

You see color bands that are horizontal. Gravity drains soap down the film, decreasing thickness from top to bottom, with constant thickness from left to right. This produces changing color bands from top to bottom, with each band of color constant from left to right.

tasks 21-22 A

a.

b. Cohesion contracts in the soap film from all directions to minimize its surface area into this particular shape.

tasks 21-22 B

A sphere encloses the maximum amount of air with a minimum surface area of soap film.

task 22

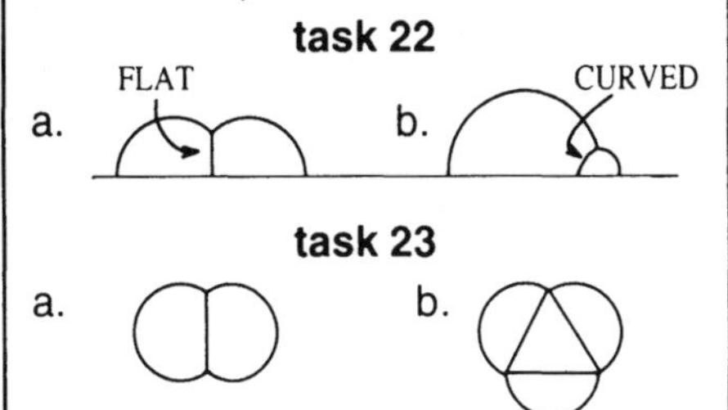

task 24

a. Bubble domes thin and break at the top because of moisture loss through drainage and evaporation.
b. You can make them larger by blowing through a larger diameter tube that holds more bubble solution.

TEACHING NOTES

For Activities 1-24

Task Objective (TO) compare the size and shape of drops of different liquids. To order these liquids according to their relative cohesive strength.

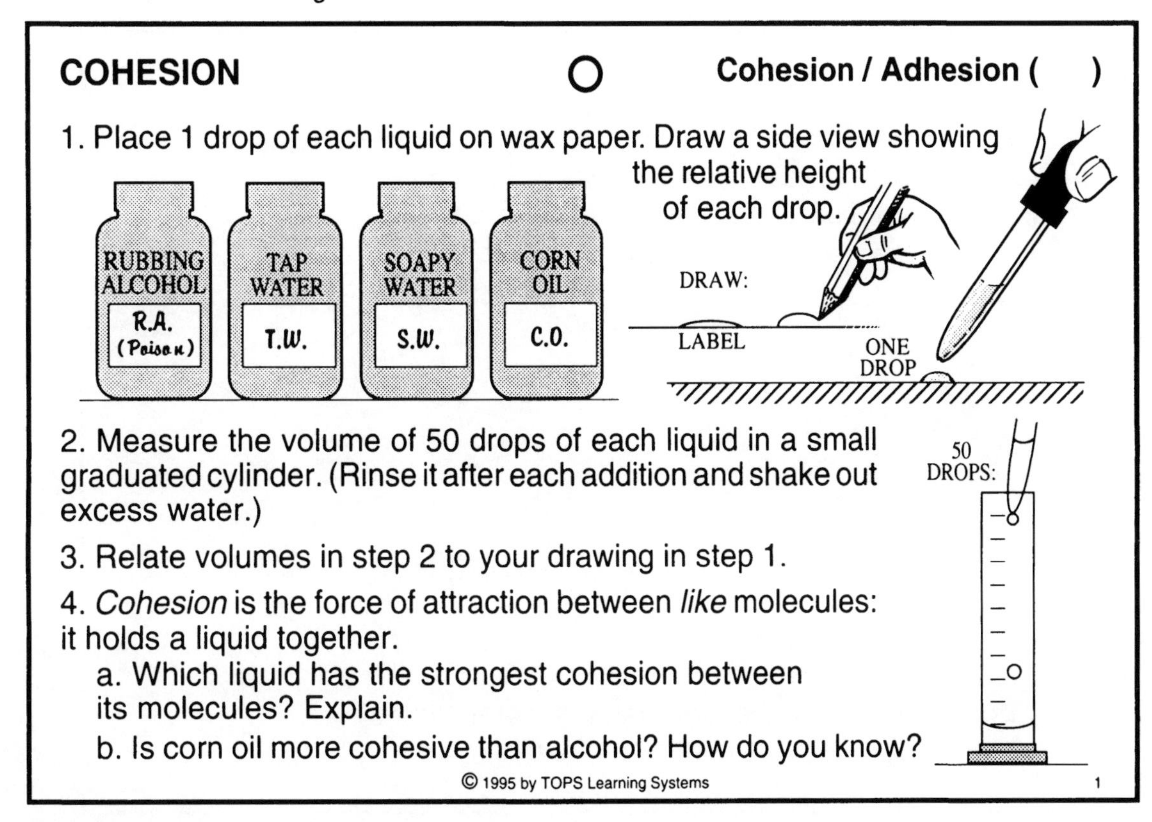

COHESION ○ **Cohesion / Adhesion (   )**

1. Place 1 drop of each liquid on wax paper. Draw a side view showing the relative height of each drop.

2. Measure the volume of 50 drops of each liquid in a small graduated cylinder. (Rinse it after each addition and shake out excess water.)

3. Relate volumes in step 2 to your drawing in step 1.

4. *Cohesion* is the force of attraction between *like* molecules: it holds a liquid together.
   a. Which liquid has the strongest cohesion between its molecules? Explain.
   b. Is corn oil more cohesive than alcohol? How do you know?

 1

Introduction

Draw an enlarged view of a 10 mL graduated cylinder on your blackboard. Count the calibrations (...2.0 mL, 2.2 mL, 2.4 mL...) to emphasize that it is divided in increments of 0.2 mL. Draw a water meniscus at various levels and ask students to read the volume.

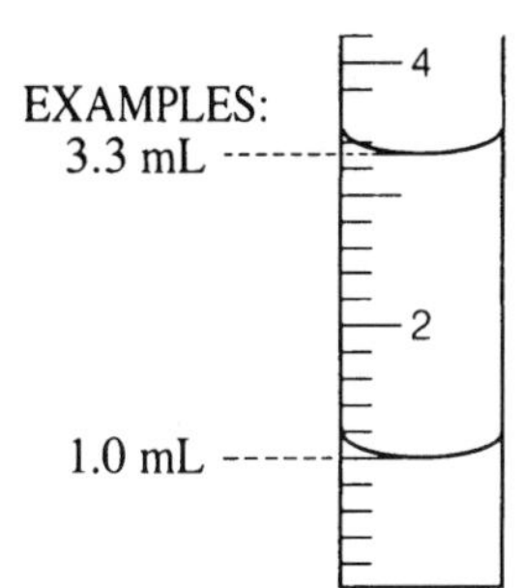

Answers / Notes

1. RUBBING ALCOHOL, TAP WATER, SOAPY WATER, CORN OIL

2. Answers will vary depending on eye dropper openings. Here is one set of results: R.A.=1.0 mL; T.W. = 2.7 mL; S.W. = 1.0 mL; C.O. = 1.6 mL

3. Liquids that heap higher on waxed paper (in step 1) have greater 50-drop volumes (in step 2).

4a. Water molecules hold together with the strongest cohesive force. Tap water forms drops that are higher and rounder, and occupy more volume than any other liquid.

4b. Yes. A drop of corn oil heaps higher on wax paper than does a drop of rubbing alcohol. Moreover, 50 drops of corn oil occupy a volume of 1.6 mL, compared with only 1.0 mL for rubbing alcohol.

Materials

☐ Four clean dropper bottles, with eye droppers of equal size. Use masking tape to label each as underlined:
- 70% Rubbing Alcohol: R.A. (poison). Purchase this in any drugstore. Tint with 1 drop blue food coloring.
- Tap Water: T.W. Rinse the bottle of all traces of soap before filling. Tint with 1 drop blue food coloring.
- Soapy Water: S.W. Add 6 drops of Joy® or Dawn® liquid detergent to the bottle and fill with tap water. (These brands work best in a bubble solution used in activities 20-24.) Tint with 1 drop blue food coloring. *Note: Though soap and detergent are chemically distinct, they have similar surfactant properties that create "soapy" solutions. For lack of a useful substitute, we use the common sense meaning of this word.*
- Corn Oil: C.O. Purchase this in any grocery store. Do *not* add food coloring.

☐ Wax paper.

☐ A 10 mL graduated cylinder. Larger capacity cylinders are *not* good substitutes.

☐ A source of water.

(TO) observe how cohesion holds water together on a penny. To recognize that soap weakens this cohesion.

HEAP O' WATER

1. Calculate how many drops of tap water add up to 1 mL. Set up a proportion using numbers from the last activity.

$$\frac{\text{X DROPS}}{\text{1.0 mL}} =$$

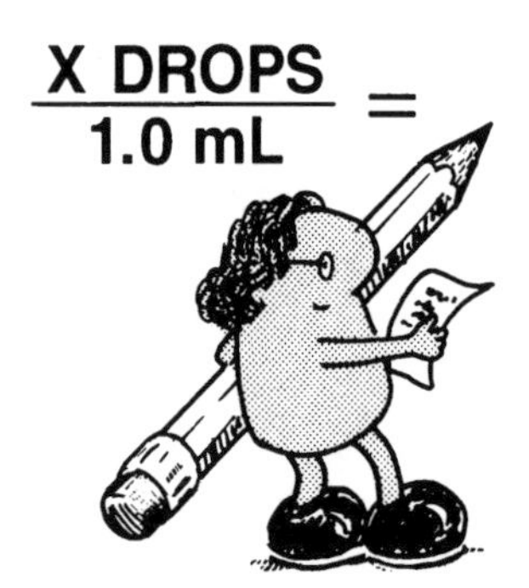

2. Try to heap 1 mL of tap water on top of a penny without spilling any over its edge. Explain your results in terms of cohesion.

3. Dry the penny. Add 1 drop *less* than a full mL of tap water to the penny without spilling.
   a. Now add 1 drop more of soapy water. What happens?
   b. How does soap affect the cohesion of water?

 2

Answers / Notes

1. In the previous activity, 50 drops of tap water was found to occupy 2.7 mL.

$$\frac{\text{x drops}}{\text{1.0 mL}} = \frac{\text{50 drops}}{\text{2.7 mL}}$$

$$x \approx 19 \text{ drops}$$

2. The penny held 19 drops of tap water (and more) without spilling over. Strong cohesive forces enabled the water drop to pull together, heaping higher and higher, until these forces were finally overcome by gravity.

Students will naturally compete with each other to heap a record number of drops on the penny. For best results, they should thoroughly dry their pennies between trials.

3a. With the addition of a single drop of soapy water, the water mound immediately collapses and spills over the edge of the penny.

3b. Soapy water dramatically reduces the cohesion of water allowing it to spill over the side of the penny. *(This overflow could not be caused by excess volume, because it was already demonstrated in step 2 that the penny has a holding capacity of at least 1 mL.)*

Materials

- ☐ Experimental results from activity 1.
- ☐ A calculator (optional).
- ☐ A penny.
- ☐ Dropper bottles of tap water and soapy water.
- ☐ A paper towel.

(TO) compare run times and trail patterns for various liquids that move down a wax paper ramp. To observe that different liquids adhere to wax paper with different strengths.

ADHESION

Cohesion / Adhesion ( )

1. Tear off a square-shaped piece of wax paper as wide as the roll, and fold to one-quarter size. Trim, tape and label the torn edge as shown:

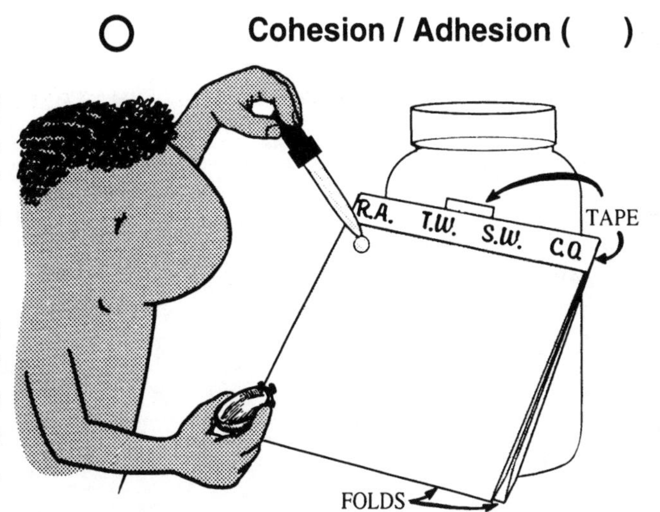

2. Tape it against a jar to form a steep ramp. Time how many seconds it takes for one drop of each liquid to move from the top of this ramp to the bottom.

3. *Adhesion* is the force of attraction between *unlike* molecules that allows the surface of one substance to stick to another. Which liquid adheres to wax paper with the greatest force? Least force? Explain.

4. Draw the trails left by rubbing alcohol and corn oil on wax paper. Relate these patterns to adhesion.

 3

Answers / Notes

2. Times will vary widely depending on the steepness and length of the wax paper slope. Tap water moves the fastest; corn oil moves the slowest. Rubbing alcohol and soapy water move at about the same speed. Here is one result:

LIQUID	RUN TIME
R.A.	5 seconds
T.W.	< 1 second
S.W.	5 seconds
C.O.	90 seconds

3. The drop of corn oil adheres most strongly to wax paper as demonstrated by its slowest travel time. The drop of tap water adheres most weakly, slipping rapidly down the wax paper without breaking up.

4. 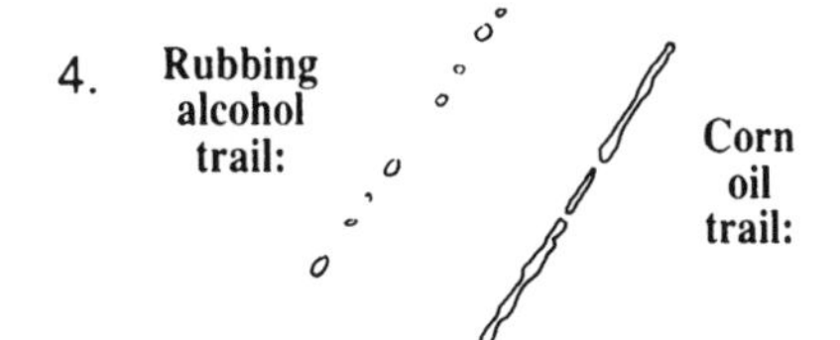

Corn oil forms a more continuous trail on wax paper than rubbing alcohol. This suggests that corn oil adheres to wax paper with greater force than rubbing alcohol.

Discussion

The soapy water drop quickly soaks through the glossy, semitransparent waxed surface of the paper and turns it an opaque white. By weakening the cohesive bonds between water molecules, soap makes water *wetter*, allowing it to quickly saturate dirty laundry and carry away the dirt.

Demonstrate this enhanced wetness by laying down a row of soapy water drops next to a row of tap water drops on wax paper. Use a paper towel to remove one drop from each row every 30 seconds.

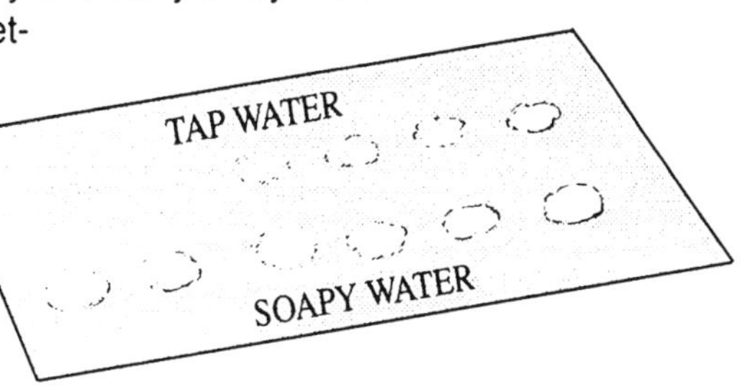

Materials

- ☐ Wax paper.
- ☐ Scissors.
- ☐ Masking tape.
- ☐ A jar or other tall object to support the wax paper ramp.
- ☐ The dropper bottles of rubbing alcohol, tap water, soapy water and corn oil from activity 1.
- ☐ A wall clock or watch with a second hand.

(TO) observe the interactions of tap water, rubbing alcohol and soapy water with corn oil and wax paper surfaces. To explain these interactions in terms of adhesion and cohesion.

CREEPY CRAWLIES O **Cohesion / Adhesion ( )**

1. Form 3 separate puddles of corn oil on wax paper, with 2 drops in each puddle.

2. Now add 1 drop of each liquid to the *edge* of a puddle. Aim well, so half of each drop touches the wax paper while the other half touches the oil.

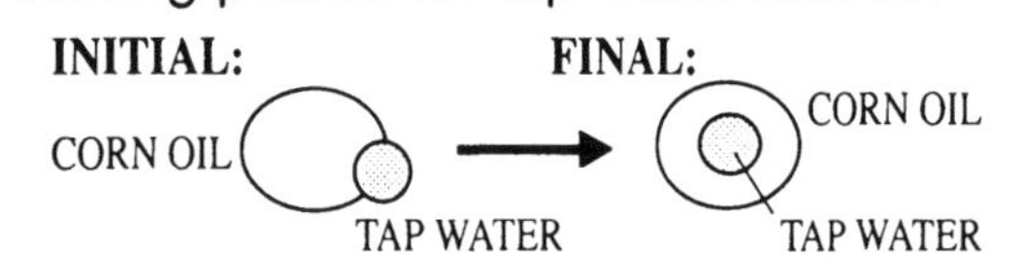

3. Here is a top view of *initial* and *final* resting places for tap water and oil.

INITIAL: **FINAL:**

CORN OIL → CORN OIL

TAP WATER TAP WATER

a. Make similar labeled drawings for all 3 puddles.
b. Explain how forces of adhesion and cohesion act on each liquid combination to rearrange it.

4. Make a 4-drop oil puddle with a 1-drop tap water "eye" in the center.

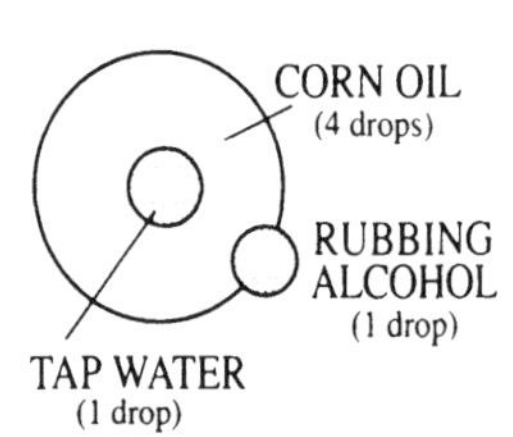

a. Add 1 drop of alcohol so it touches *only* the edge of the oil puddle but *not* the middle eye.
b. Describe the delayed reaction in terms of cohesion and adhesion.

 4

Answers / Notes

3.

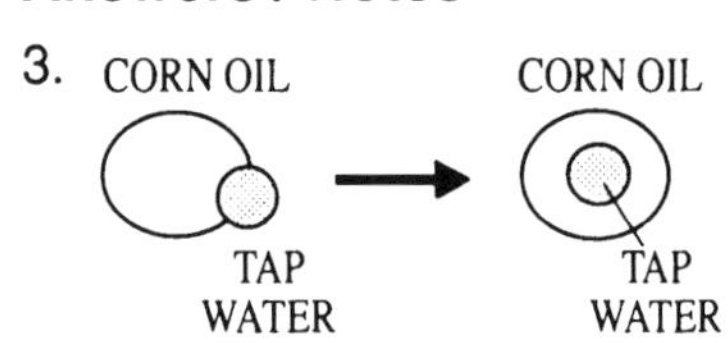

Tap water adheres with greater force to the surface of corn oil than to wax paper. This force difference pulls the drop onto the corn oil puddle, away from the wax paper. Strong cohesive forces in the tap water prevent the drop from breaking apart as it is pulled along.

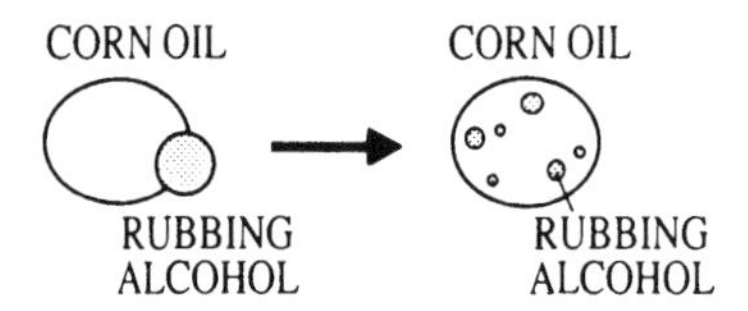

Rubbing alcohol adheres with greater force to the surface of corn oil than to wax paper. This force difference pulls the drop onto the oil puddle, away from the wax paper. The cohesive forces in the alcohol are too weak to prevent the drop from spreading over the oil puddle in a thin film. After 3 or 4 minutes, weak cohesive forces in the alcohol rapidly recondense this film into several small droplets. *(Eventually, these droplets slide off the edge of the oil puddle and soak into the paper.)*

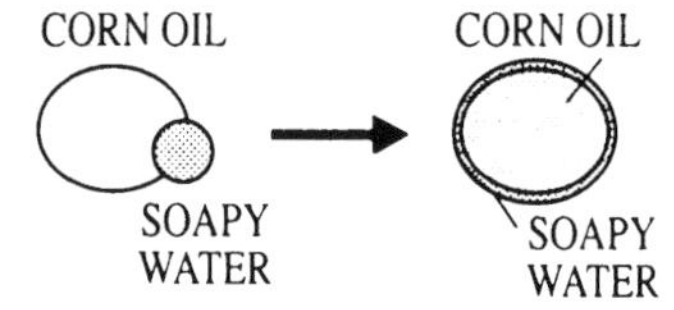

Soapy water adheres with greater force to wax paper than to the surface of the oil. With only very weak cohesive forces holding it together, the water spreads over the wax paper surface and soon soaks in. Adhesive forces are strong enough to draw it completely around and under the corn oil puddle.

4b. Adhesion with corn oil overcomes cohesion in the rubbing alcohol drop and adhesion with wax paper. The alcohol advances as a thin film over the corn oil until it touches the tap water "eye" and mixes with it vigorously. This reduces cohesion in the "eye," causing it to flatten and spread over a wider area.

Materials

- ☐ Wax paper.
- ☐ Dropper bottles of rubbing alcohol, tap water, soapy water and corn oil.

(TO) model the structure of water. To understand its cohesiveness in terms of electrostatic attraction.

H_2O ○ Cohesion / Adhesion (   )

1. Trim the Water Molecules cutout around its outer solid line. Cut along the thin dashed lines as follows:
   a. Fold along center line A. Cut along the 7 dashed lines to each *molecule*.
   b. Repeat along lines B and C.
   c. Cut in from edges D and E.

CUT
CUT
FOLD LINE A

2. Stretch these molecules into a honeycomb: twist the left one up, the right one down, and pull gently apart.
   a. Fold the tabs over the rim of a paper plate and tape in place.
   b. Label your model like this:

PULL & TAPE
PULL & TAPE
LABEL

WATER
(127 million times actual size)

3. One oxygen atom and two hydrogen atoms form a molecule of water joined by two covalent bonds. Water molecules have electron-rich areas and electron-poor areas that mutually attract to form hydrogen bonds.
   a. Copy the above description. Use your model to point out each underlined feature to a friend.
   b. Which bonds are stronger (closer together)? Which bonds are weaker (farther apart)?

 5

Introduction

• Fill a glass with water. Pour half of it down the sink, then half of the remainder, then half of that, etc., until you are left with 1 drop. Start chopping the drop in half with an imaginary knife. How small can you go and still have water? (Down to 1 molecule. Divide this, and you get 1 oxygen atom and 2 hydrogen atoms.)

• Hand out the Water Molecules cutout. Identify a *molecule* of H_2O with its two atoms of *hydrogen* and one atom of *oxygen*. These three atoms share electrons, which bind them tightly in *covalent bonds*. The distribution of electrons creates electron-poor areas of positive charge and electron-rich areas of negative charge (charges shown in black are in front, and white, behind). Like charges repel and opposite charges attract, so these molecules align as modeled, forming electrostatic *hydrogen bonds* between molecules.

Answers / Notes

HYDROGEN BONDS:

2a. *The two black (front) charges and two white (back) charges suggest, in 3 dimensions, an equidistant distribution of positive and negative charge around each water molecule. If you imagine one molecule at the center of a tetrahedron, then its 4 hydrogen bonds radiate outward toward each vertex. Our paper model unavoidably flattens this tetrahedral orientation to some extent.*

2b. A masking tape label should read "Water (127 million times actual size)."

3a. *Students should copy the descriptive paragraph, then explain it to each other.*

3b. Covalent bonds are stronger and closer together. *(These are the bonds that break when water is split by electrolysis into hydrogen and oxygen gas.)* Hydrogen bonds are weaker and farther apart. *(These are the bonds that break when liquid water evaporates to a gas.)*

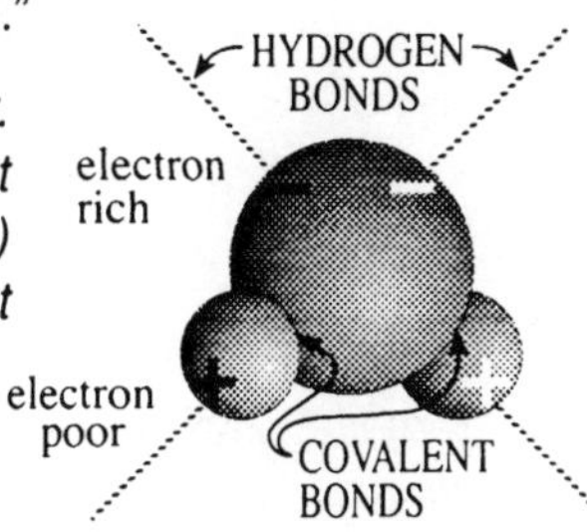

Materials

☐ The Water Molecules cutout. Photocopy this from the supplementary page at the back of this book.
☐ Scissors.
☐ A standard paper plate, 9 inches (23 cm) in diameter.
☐ Masking tape.

(TO) observe how small particles on the surface of water move in response to breaking hydrogen bonds.

ZIPPERS ○ **Cohesion / Adhesion ( )**

1. Untape your Water Molecules model from the paper plate, and pull gently on the tabs. How does this model explain the source of cohesion in real water?

2. Fill a clean glass with tap water and sprinkle pepper on its surface.

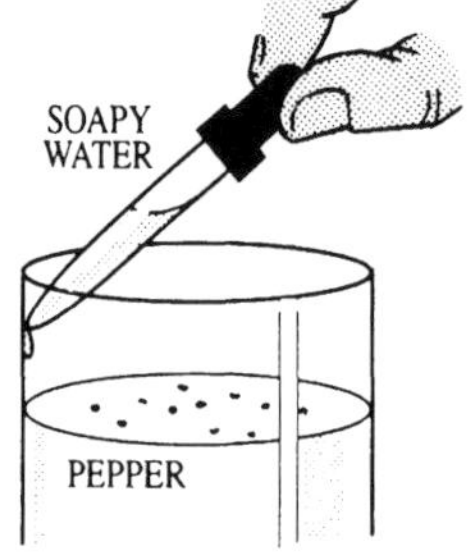

a. What happens when you drip 1 drop of soapy water down the inner wall of the glass?

b. What happens when you stretch your Water Molecules model, then release one side?

c. Explain how 2b models 2a.

3. Rinse your glass and set it in a bowl. Fill it higher than the brim with fresh tap water.

a. Scrape a *tiny* speck off a bar of soap with a pin. Drop it on the water, and observe carefully.

b. Where does this soap speck get its energy?

4. Add a few crumbs of solid camphor (the smaller the better) to a clean glass of tap water. Compare and contrast camphor specks with soap specks.

 6

Answers / Notes

1. Water is strongly cohesive because its molecules form a network of hydrogen bonds, points of electrostatic attraction between electron-poor hydrogen atoms and electron-rich oxygen atoms.

2a. The pepper rapidly retreats from the point where the soapy drop first touches the water's surface.

2b. The Water Molecules move closer together when you release one side.

2c. Releasing one side of the Water Molecules model simulates the breaking of hydrogen bonds. Unbroken bonds pull back in response *(much like one team in a tug-of-war falls backward when the other side drops the rope).* In a similar manner, soap breaks real hydrogen bonds in real water. This creates an unbalanced contraction of hydrogen bonds which pull surface molecules away from the break, and the floating pepper with them.

3a. The tiny soap speck twirls and spins on the surface of the water for a few seconds, then slows down. *(If this doesn't happen, refill the glass with fresh tap water and drop in a* smaller *speck of soap.)*

3b. The speck of soap breaks nearby hydrogen bonds wherever it moves, and gets carried along by unbroken hydrogen bonds that contract in response. *(This process slows to a halt as all hydrogen bonds in the immediate area of the soap speck are broken up.)*

4. Camphor specks energetically zip over the surface of the water, like the soap speck, only for a much longer time. *(The camphor is apparently much less soluble in water, breaking far fewer hydrogen bonds as it moves about. As the density of unbroken hydrogen bonds decreases, these specks gradually slow down. Over time they dissolve.)*

Materials

- ☐ The Water Molecule model previously constructed.
- ☐ A glass, bowl and jar.
- ☐ Tap water.
- ☐ A shaker of finely ground pepper.
- ☐ A dropper bottle of soapy water.
- ☐ A pin and a bar of soap.
- ☐ A piece of solid camphor, synthetic or natural. Ask for this over-the-counter cold remedy at your local drug store. After opening, store it in a closed baby food jar. It is volatile and highly aromatic. (If unavailable in one store, try another. Camphor's action in water is fascinating. As a last resort, skip step 4.)

(TO) recognize that water cohesion creates an unbroken skin-like surface strong enough to support a pin. To distinguish between this phenomenon and floating.

SURFACE TENSION ○ **Cohesion / Adhesion ( )**

1. Fill a bowl half full of tap water. Float a toothpick in the bowl.

2. Stick 2 pin "arms" in a bit of styrofoam as shown, to cradle a third pin sideways.

3. Gently lower the cradled pin onto the water beside the toothpick. *Surface tension* (a skin-like property of water) prevents the pin from sinking.

a. Prove that the pin doesn't float like a toothpick.
b. With your Water Molecules model and a pencil, model how real water supports the pin.

4. Support the pin with surface tension as before, next to the floating toothpick. Sprinkle pepper over the water.

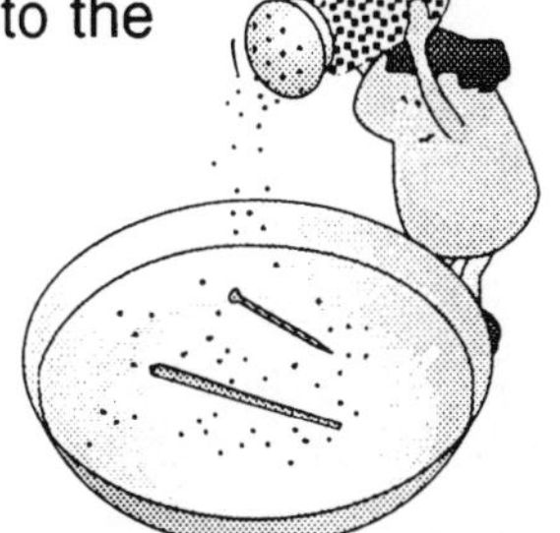

a. A pattern in the pepper tells you that the pin rests on water, while the toothpick floats in water. Explain. Illustrate your answer with a drawing.
b. What will happen if you squeeze a dropper full of soapy water into the bowl? Make a reasoned prediction, then test it.

 7

Answers / Notes

3. *When you use the pin "arms" to place the pin, it should easily rest on top of the water. If it sinks, rinse the bowl of any residual soap, dry off the pin and try again.*

3a. Students should push both objects under the water: the pin sinks to the bottom; the toothpick rises again to the surface.

3b. Students should rest a pencil on the "hydrogen bonds" of their stretched-out Water Molecule model. As long as these bonds remain unbroken, the pencil doesn't fall through. Hydrogen bonds in real water create a similar network of bonds that are strong enough to support a pin.

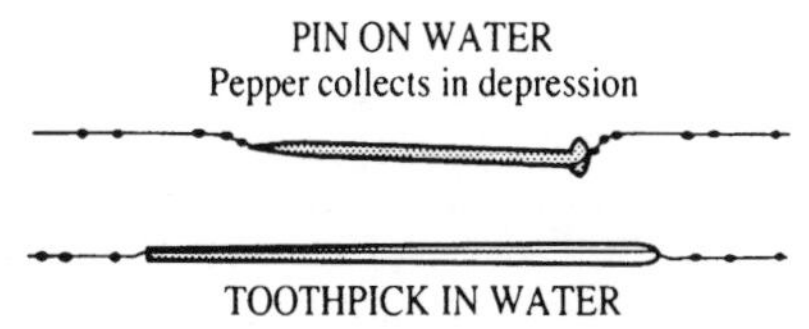

4a. Pepper collects in higher concentrations around the perimeter of the pin. It falls into the surface depression created by the pin pressing on the surface of the water from above. Pepper does not collect around the perimeter of the toothpick, because there is no depression. The toothpick has broken through the hydrogen bonds on the surface to float in the water.

4b. Prediction: The soapy water will break hydrogen bonds on contact to open a "hole" in the water's surface skin. Unbalanced surface tension from hydrogen bonds that remain strong will sweep the pepper and pin away from this surface break. As hydrogen bonds under the pin also weaken and break, the pin will sink.

Students should evaluate their predictions. *(The pin may not sink immediately in cold water, but will do so after a few seconds.)*

Materials

☐ A bowl of tap water.
☐ Three aluminum pins. Steel pins tend to magnetize and "stick" to the cradling pin arms. This complicates a quick, clean surface release.
☐ A chip of styrofoam, 2 cm or greater across, from a cup, meat tray, etc.
☐ A toothpick.
☐ Pepper. Use a pepper shaker if available. Otherwise, use your fingers to sprinkle a pinch over the water.
☐ A dropper bottle of soapy water.

(TO) explain interactions of water, styrofoam and string in terms of adhesion, cohesion and surface tension.

GROOVY

○ **Cohesion / Adhesion (   )**

1. Press a dent into the lip of a styrofoam cup with the edge of 2 pennies. Cut a narrow V-groove into the cup, as wide as this dent at the lip, and about 1.5 penny diameters deep.
2. Tint a large jar of water with a drop of blue food coloring. Stand your V-cup in a bowl and gently fill it with blue water until it spills out the groove.
   a. How high can water rise in the cup? Describe how it spills down the side.
   b. Explain each observation in terms of surface tension, adhesion, and cohesion.
3. Spiral a blue stream around the white cup. Explain how each tool uses adhesion and cohesion to advantage:
   a. Use a *straw* to guide the stream in a spiral half-way around the cup.
   b. Rechannel the stream with a *wet string.* Spiral the stream in a full circle.

 8

Answers / Notes

2a. Water rises above the V-groove, nearly to the rim of the cup, without spilling. Then it bursts out near the bottom of the groove in a narrow stream that snakes left and right down the outside of the cup, occasionally changing course.

2b. Surface tension holds the water back like a dam as it rises high into the V-groove. Pressure builds near the bottom of this groove, finally pushing water molecules through the restraining "skin." Adhesion between water and styrofoam prevent this outflow from falling free. Instead, it runs down the outside of the cup in a meandering stream. Cohesion between water molecules pulls new water behind old in a continuous stream. Because adhesion between the water and styrofoam is relatively weak, the flow is heavy enough to occasionally break the stream free of established channels to follow a new path.

3a. Blowing gently against the side of the stream breaks weak adhesive bonds between water and styrofoam, redirecting the stream to flow diagonally in a half spiral to the back of the cup. Cohesion drags additional water along the same channel, maintaining a continuous flow. *(If you blow too hard, you break the cohesive forces holding the stream together. If you redirect the stream in a diagonal that is too shallow, gravity redirects the stream downward.)*

3b. Wedge an end of the wet string in the V- groove and wind it in a full spiral around the cup. Adhesion between water, string and styrofoam hold the string in place against the cup. Fill the cup above the notch, and the water spirals all the way around the cup, channeled by cohesion and adhesion along the wet string.

Materials

- ☐ A styrofoam cup.
- ☐ Two pennies.
- ☐ Scissors.
- ☐ A large jar of water.
- ☐ Blue food coloring in a dropper dispenser.
- ☐ A bowl to catch water.
- ☐ A plastic straw.
- ☐ A piece of string at least 20 cm long.

(TO) observe how water moves between glass slides by capillary action. To understand how adhesion and cohesion drive this process.

CAPILLARY ACTION O Cohesion / Adhesion ( )

1. Fully wax 1 side each of two glass slides with the *side* of a candle. Face the waxed sides together, offsetting them as shown.
2. Position a pair of unwaxed slides like the first.
3. Place 1 drop of tap water on the exposed "lip" of each pair. Does water adhere with greater force to glass or to wax? Explain.

WAXED SLIDES
UNWAXED SLIDES
WATER DROP

4. Move your head to see room light reflected on the unwaxed pair. Keeping this light in view, move the top slide over to just touch the water drop: draw and describe what you see.
   a. Water moves between the slides by a process called *capillary action.* Explain how adhesion and cohesion drive capillary action.
   b. Pick up the top slide. Why does the bottom slide stick to it?
5. Repeat step 4 for the waxed slides. Can you observe capillary action? Why?
6. Save both pairs of slides for the next activity.

MOVE THE TOP SLIDE TO TOUCH THE WATER.
REFLECTED LIGHT

Answers / Notes

3. The drop on the glass spreads much flatter and wider than the drop on the wax. This suggests that water adheres more strongly to glass than to wax.

4. Water creeps from the drop into the narrow interface between the wax-free slides. Except for isolated air bubbles, it spreads between the entire interface.

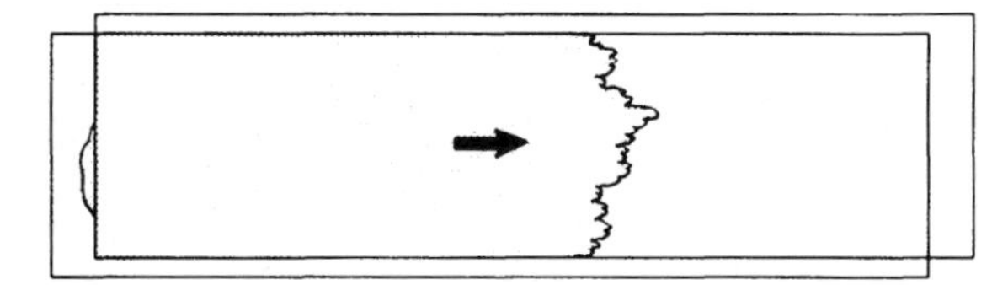

4a. On contact, adhesion with the glass pulls water from the drop into the narrow space between the slides. Cohesion draws additional water behind it, creating an advancing front that continues to move as long as the original drop lasts.

4b. The slides are held together by adhesion (the attraction of water to glass), and cohesion (the attraction of water to itself). *Because these forces are applied over such a wide surface area, the effect is unusually strong.*

5. Little or no capillary action occurs. The drop does not move into the narrow space between the waxed slides because the force of adhesion between wax and water is too weak to drag water out of its cohesive drop.

Materials

- ☐ Four glass microscope slides.
- ☐ A candle with smooth sides. Or substitute a wax crayon of any color with its paper covering removed.
- ☐ A dropper bottle of tap water.

(TO) understand why capillary action increases as water is squeezed within narrower space between glass.

WALL 'O WATER

Cohesion / Adhesion ( )

1. Get both pairs of slides from the last activity.

2. Sandwich a paper clip between the unwaxed slides as shown, so it rests in an upper corner while you hold the opposite upper corner.

3. Dip the bottom edge into a bowl of water. Make a drawing to show how high water rises up between the slides.

4. Why does the water rise to form a "wall?" Why does this wall curve upward?

5. Repeat this experiment with your pair of waxed slides. Do you observe capillary action? Why?

 10

Answers / Notes

3.

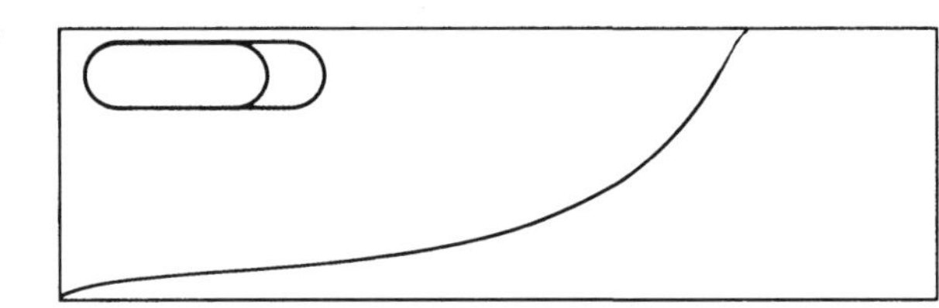

4. Water rises between the slides because of capillary action. It is attracted to the glass by adhesion. These adhesive forces act with great force between the slides, because here the water is able to contact the glass on both sides. As adhesion pulls the water up, cohesion drags additional water up from behind to form a continuous wall.

Capillary action pulls up water until its upward force is balanced by the weight of the water pushing down. This forms a wall of water that is lowest and thickest below the paper clip, where the glass slides are most widely separated. As the glass surfaces are pinched closer together, the same capillary attraction works on a thinner wall of water (with less weight) to raise it higher. This water wall continues to curve upward as it narrows until it reaches the top edges of the glass.

5. Little or no capillary action occurs between the waxed surfaces. The force of adhesion between wax and water is too weak to lift the water.

Materials

- ☐ The pair of waxed and pair of unwaxed microscope slides from the last activity.
- ☐ A paper clip.
- ☐ A bowl of water.

(TO) observe and explain surface interactions between glass and water in terms of cohesion, adhesion, surface tension and capillary action.

WATER LINES  Cohesion / Adhesion (   )

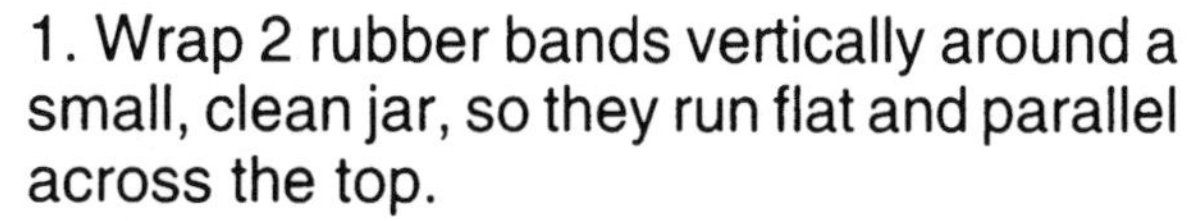

1. Wrap 2 rubber bands vertically around a small, clean jar, so they run flat and parallel across the top.
2. Get a clean glass eyedropper tube with the rubber bulb removed. Stick it between the rubber bands, tip pointed up.
3. Set this jar in a bowl, and fill it brimful with tap water. Gently snap the tip of the tube with your finger to dislodge any water adhering inside the tip.
4. Copy a large side view of the jar and tube like this. Accurately draw how all curving surfaces of water meet the glass.
5. Note where different shapes in the water's surface are influenced by cohesion, adhesion, surface tension and capillary action.
6. How do you think your drawing would change if the glass tube were thinner?

 11

Answers / Notes

3. *The pointed tip of the glass tube must be free of water. Otherwise, increased pressure inside the tube may prevent water from rising by capillary action.*

4-5.

Adhesion of water to the inner walls of the glass tube create a curved meniscus. Cohesion holds this meniscus together.

Surface tension (cohesion) holds the water together, allowing it to pile up at the lip of the glass without running over.

Capillary action, a combination of adhesion and cohesion, draws water up inside the tube.

Adhesion attracts water and glass.

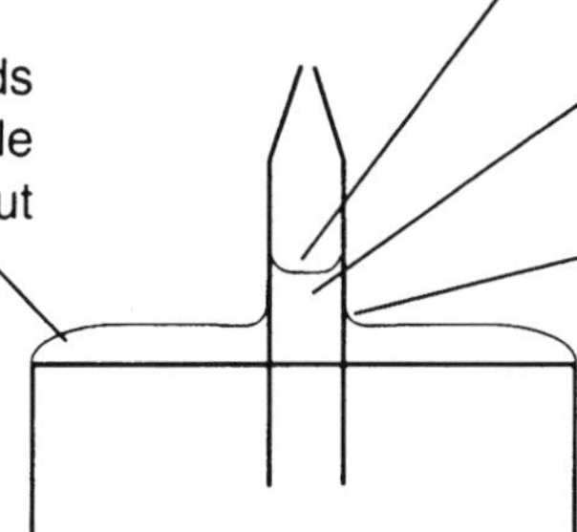

6. If the glass tube were thinner, capillary action would pull up water to a greater height before it was balanced by the weight of the water pulling down.

Materials

- ☐ A baby food jar or equivalent.
- ☐ Two narrow rubber bands.
- ☐ An eye dropper tube with the rubber bulb removed.
- ☐ A bowl to catch water.
- ☐ A jar of water.

(TO) study capillary action in a paper towel. To relate this process to the transpiration of water through trees.

CAPILLARY PATHWAYS  Cohesion / Adhesion ( )

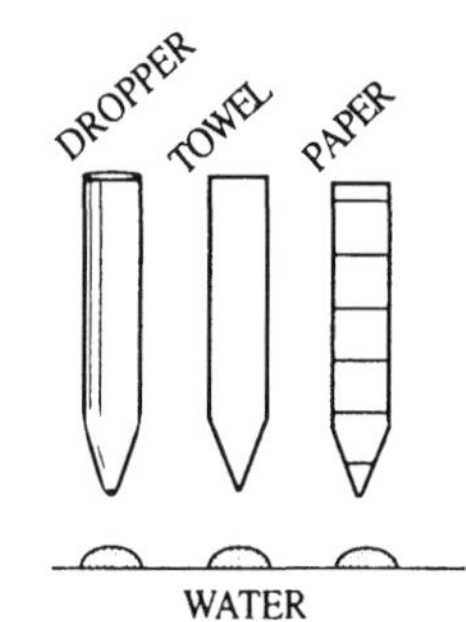

1. Get a clean glass eye dropper with the rubber bulb removed. Cut a strip of soft paper towel and a strip of notebook paper into a similar shape.

2. Put 3 separate drops of tap water on wax paper. Touch the point of each "dropper" to one of these drops.
   a. Which droppers have capillary action? Which do not?
   b. Examine the paper droppers with a hand lens. Propose a theory to explain why only one has capillary action.

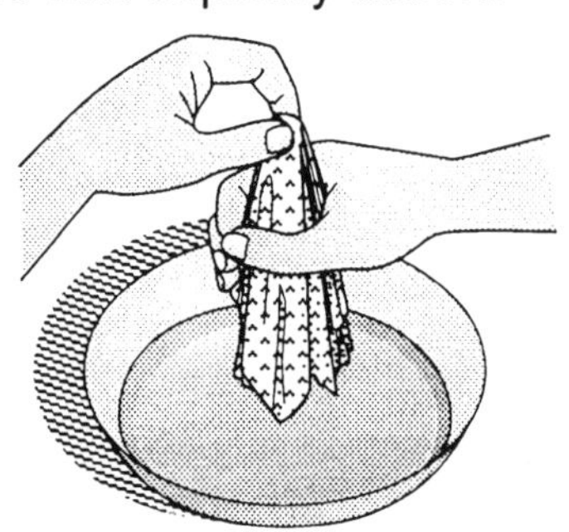

3. Put water in a bowl, 1 cm deep, and add a drop of blue food coloring.
   a. Pinch a paper towel in the middle. Squeeze the rest into a narrow cone, and stand it in the water.
   b. What happens?

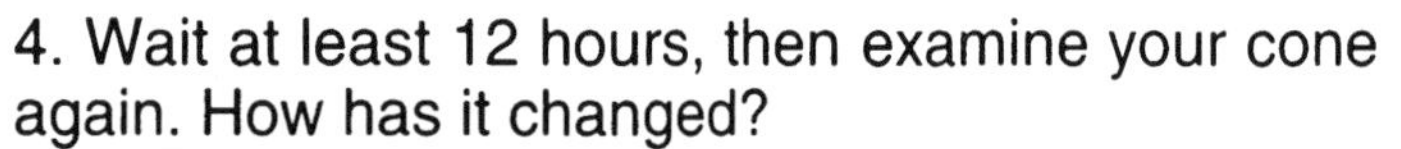

4. Wait at least 12 hours, then examine your cone again. How has it changed?
   a. Propose a theory to explain your observations.

   b. Explain how this models *transpiration*, how water moves through trees?

 12

Introduction

If your students are not already familiar with the idea that liquids evaporate, try this quick review:
a. Fill a glass with water and set it on your desk. Discuss its fate over the next few months if no one disturbs it.
b. Summarize your discussion with this blackboard equation: $\text{liquid} \xrightarrow[\text{evaporates}]{} \text{gas}$

Answers / Notes

2a. Both the glass dropper and the paper towel pull up the water drops by capillary action. Water adheres to the notebook paper, but isn't pulled up.

2b. The dropper cut from towel is composed of 2 layers of soft, textured paper with numerous tiny holes and air spaces. Adhesion pulls water inside on contact, where it then advances in tiny cohesive streams through a capillary network of cracks, holes and crevices. Notebook paper, by contrast, is smooth and dense. Water adheres only to its surface. Its dense interior provides no capillary pathways to advance the water.

3b. Water advances up the base of the cone by capillary action.

4. The entire cone is now wet. *(Under conditions of low humidity the top of the cone might be dry.)* Its outer surface is stained by blue dye, especially on the ridge folds. The interior surfaces remain uncolored.

4a. Water soaks into all parts of the cone by capillary action, and evaporates off its outer surfaces, especially at the ridge folds. Dye molecules carried by water through the cone do not evaporate. These molecules move into areas of high evaporation and accumulate as concentrations of blue stain.

4b. The base of the cone models the roots of a tree; its interior layers represent the trunk and branches; its outer surface stains model the leaves. By capillary action this "tree" draws water out of the ground, upward through its trunk and branches, and out into its leaves. Evaporation from the leaves maintains this flow throughout its vast network of capillaries.

Extension

Build a capillary siphon, with a jar of water, an empty jar, and a rolled paper towel. Predict the water levels in each jar after 48 hours.

Materials

- ☐ A glass eye dropper tube.
- ☐ A soft, absorbent paper towel.
- ☐ A piece of notebook paper.
- ☐ Scissors.
- ☐ Dropper bottles of tap water and food coloring.
- ☐ Wax paper.
- ☐ A hand lens.
- ☐ A bowl or saucer of tap water.
- ☐ A meter stick.

(TO) separate ink spots into their component colors using chromatography.

CHROMATOGRAPHY ○ Cohesion / Adhesion ( )

1. Loop a rubber band "clothesline" across the mouth of a small dry jar.

2. Cut 2 strips, about 1 cm wide, from the white margin of a newspaper. Fold these as shown so they hang on your line and just touch the bottom of the jar.

3. Use washable markers to draw horizontal lines 1 cm above the bottom of each strip, one Black and one Brown. Label the tops with the colors you applied.

4. Hang these strips so they don't touch each other or the jar's sides. With an eyedropper, add clear tap water to wet only the *uncolored* "foot" of each strip.

5. Set up additional jars with Red, Orange, Yellow, Green, Blue and Purple.

6. Remove the strips before the colors reach the top of the jar. Tape them to your assignment sheet with writing space between them.
   a. Label all the colors you see in each strip.
   b. Which markers seem to contain a *mixture* of dyes? *Pure* dyes?

 13

Answers / Notes

4. *Students should dribble perhaps 6 droppers of clear tap water (not tinted) down the side of the jar, just covering the bottom. Colored lines that inadvertently touch the waterline will not migrate. They must be* above *water level to be transported by capillary flow.*

5. *It is possible to hang three strips per jar, but more difficult to keep them from touching each other or the sides of the jar.*

6. *Allow the colors to separate by capillary action for at least 10 minutes; up to an hour if time permits. Always remove each strip before water reaches the rubber band, or moisture will straighten the fold and the strip will fall off the line. If you substitute filter paper for newsprint, these colors will migrate much faster, reaching the top of the jar in less than 10 minutes.*

6a. *Assign at least 1 color "expert" to each lab group (boys, in particular, are prone to red/green color blindness). Our ink colors separated like this, with subtle transitions in between that we have not labeled:*

Bk	Br	Re	Or	Ye	Gr	Bl	Pu
purple	orange	red	orange	yellow	yellow	blue	red-purple
blue	green				green		
green	pink						
brown	orange				blue	pink	blue

6b. *In theory, red, yellow and blue are primary colors that make up all the rest: green, purple and orange are secondary colors composed of two primaries; black and brown are tertiaries made from all three primaries. In practice, our blue marker contained a touch of pink; our orange marker failed to resolve into red and yellow, and could be a pure dye, though evidence in the next activity suggests it is a mixture.*
Mixed Dyes: black, brown, green, purple, blue(?)
Pure Dyes: red, yellow, orange(?)

Materials

☐ Newspaper that supports capillary action. Test your source in advance by holding the end of a strip in water. After 1 minute water should have advanced at least 1 cm. If your paper is slower, test a cheaper, less dense grade of newsprint. For faster color separations that are somewhat less complete, you can substitute white, lab grade filter paper. Highly absorbent paper products like napkins or toilet paper don't work at all.
☐ Scissors.
☐ Four baby food jars and 4 thin rubber bands.
☐ A small jar of clear tap water with an eyedropper.
☐ A set of washable colored markers. The separations at left were based on Crayola® Classic markers. Other brands may give different results.

(TO) develop a color mixing chart based on color chromatograms. To model why some dye combinations separate by capillary action better than others.

MORE CHROMATOGRAPHY ○ Cohesion / Adhesion ( )

RED & YELLOW RED & BLUE BLUE & YELLOW

1. Superimpose these colored lines, one on top of the other, on strips of newspaper. Separate these pairs by chromatography as before.

BLUE RED YELLOW

2. Wait at least 10 minutes for the colors to separate, then tape the strips to your assignment sheet. Label all colors.
   a. Red, yellow and blue combine to create other colors. Study these and other chromatograms (color strips) to complete this color chart.
   b. Which dye molecules (red, blue or yellow) move slowest? Explain.
3. Cut into a piece of scratch paper about 1/5 of its length. Overlap and tape to form a scoop.

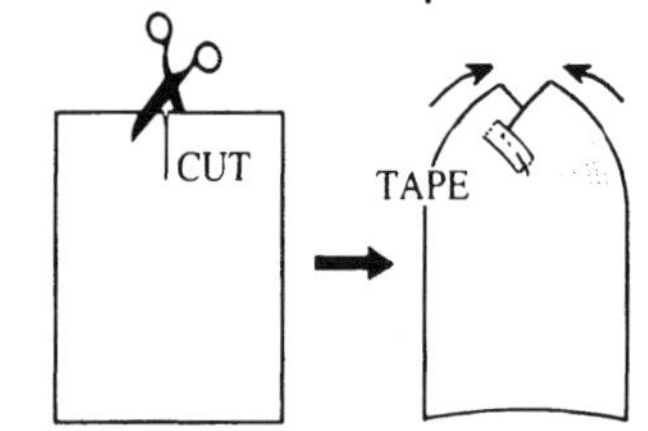

   a. Collect a pinch each of salt and pepper in the back of the scoop. Hold it at an angle and tap the end with your pencil to move these grains down its length.
   b. How do salt and pepper model the dye molecules moving through a chromatogram by capillary action?
   c. Why do some color combinations separate better than others?

SALT & PEPPER

 14

Answers / Notes

1. *Students should separate each pair of colors in small jars looped with rubber bands as in the previous activity.*

2. *Results may vary depending on the particular dyes used to manufacture your brand of markers.*

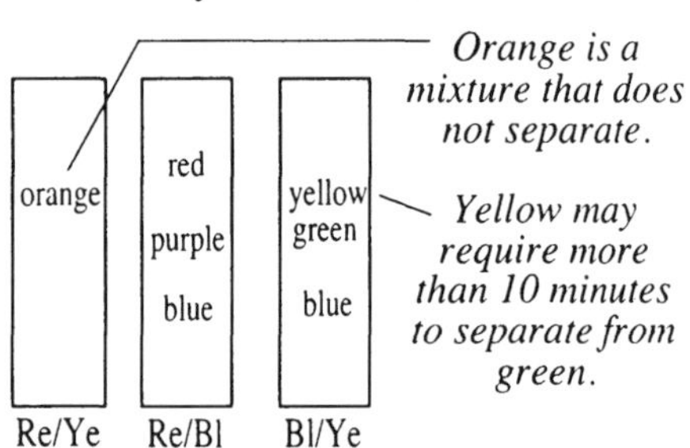

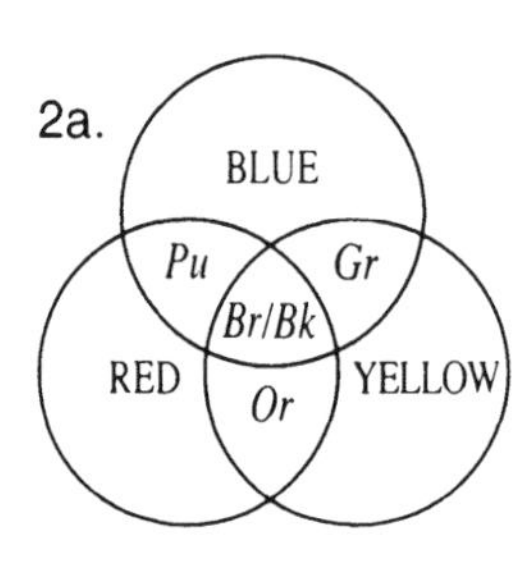

2a.

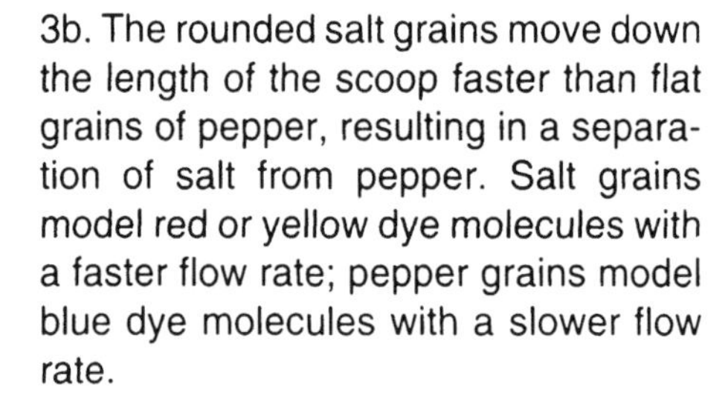

2b. Blue dye molecules moved the slowest, trailing in both the red/blue and blue/yellow chromatograms.

3b. The rounded salt grains move down the length of the scoop faster than flat grains of pepper, resulting in a separation of salt from pepper. Salt grains model red or yellow dye molecules with a faster flow rate; pepper grains model blue dye molecules with a slower flow rate.

3c. Dyes with different flow rates (Bl/Re or Bl/Ye) quickly separate while dyes with similar flow rates (Re/Ye) remain together. *(The flow rate of each particular dye is determined by its solubility in water and its adhesion to newsprint.)*

Extension

Create a radial chromatogram:

a. Cut a small hole in the middle of a circle of filter paper. Draw a black circle around it using a washable marker.

b. Poke cotton through the hole. Dip this wick in a jar of water while the filter paper rests on the rim.

c. Allow capillary action to carry the black dye outward. Stop before any color reaches the outer edge of the circle.

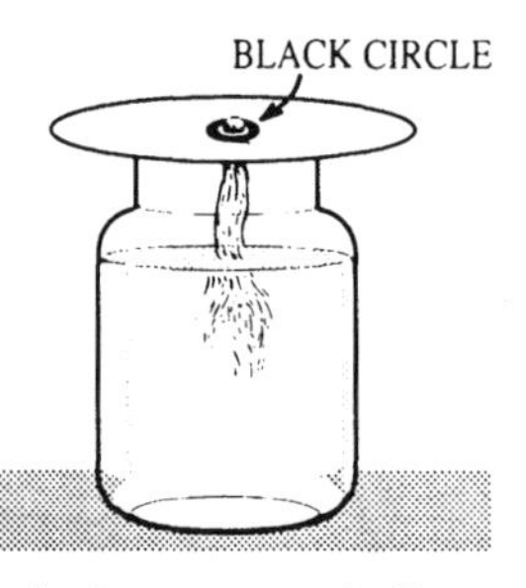

Materials

- ☐ Washable markers and chromatograms from the last activity.
- ☐ Newsprint.
- ☐ Scissors.
- ☐ Baby food jars.
- ☐ Thin rubber bands.
- ☐ Clear tape.
- ☐ Scratch paper (8 $1/2$ x 11 inches).
- ☐ Salt and pepper.

(TO) form an oil slick on water with a drop of corn oil, and disperse it with a drop of soapy water. To understand the molecular forces at play.

THICK SLICK O **Cohesion / Adhesion ( )**

1. Put a bowl of tap water near a window where you can see daylight reflected on the water.

2. Let the water calm, then place 1 drop of corn oil on the surface. Watch the oil spread over the water in the reflected light.

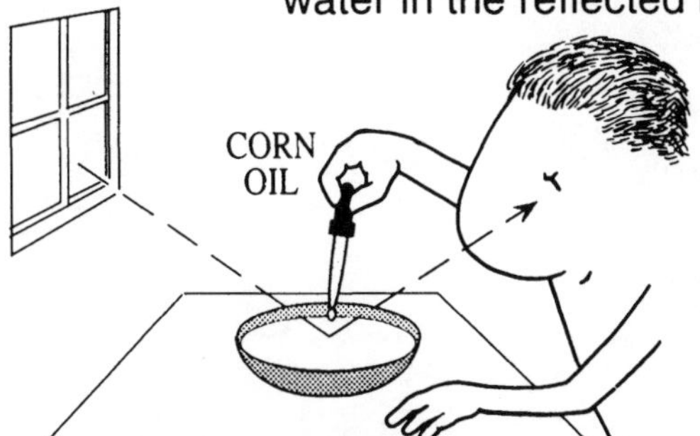

a. Draw pictures of the oil slick. How does it change over time?

b. Which is stronger, the cohesion of the corn oil, or its adhesion to the surface "skin" of water? Explain.

3. Add 1 drop of soapy water to the slick while watching it in reflected light.

a. What happened to the oil slick?

b. Did the cohesion of corn oil just get stronger, or the surface "skin" of the water just get weaker? Explain.

c. Cleanup crews often spray soapy detergents on oil spills. Explain how this makes their job easier.

 15

Answers / Notes

2a. The single drop of oil spreads out in a wide circle. Over time, holes open up within the spill, giving it a "Swiss cheese" appearance.

NEW SPILL:

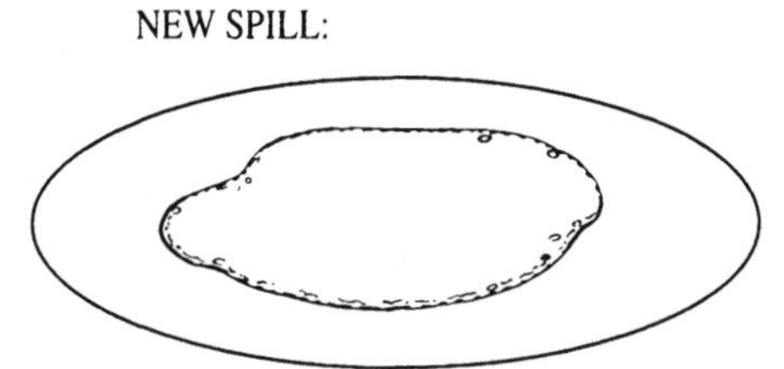

AFTER SEVERAL MINUTES:

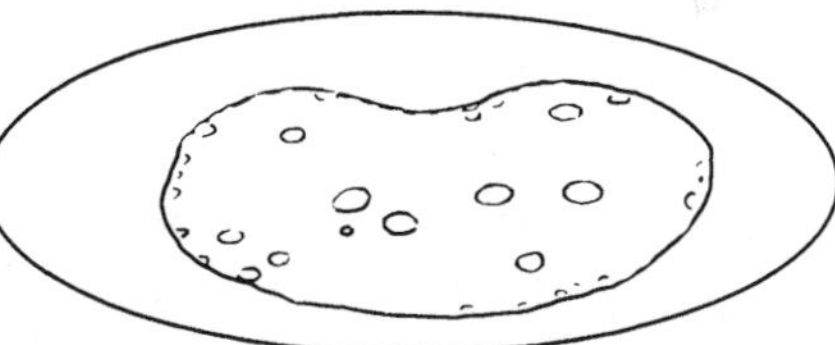

2b. The adhesion of corn oil to the surface skin of water is stronger than cohesive forces that hold the corn oil drop together. Adhesion attracts the oil drop in all directions across the surface of the water, spreading it over a large, flat circle. *(Corn oil apparently reduces the surface tension of water over time. Cohesive forces pull away from weakening points of adhesion under the slick to open up numerous holes.)*

3a. With just 1 added drop of soapy water, the oil slick breaks up and contracts into numerous small drops.

3b. The soapy water reduced surface tension in the water. Without a surface "skin" on which to adhere, cohesive forces in the flat oil slick pull it back into tiny round drops of vastly reduced surface area.

3c. These detergents reduce the surface tension of the water, allowing cohesive forces in the widely dispersed oil slick to pull the oil into more compact drops that are easier to skim away.

Extension

Set a clean bowl on an overflow plate, fill it brimful with water, then add 1 drop of corn oil. Try to clean up this "oil spill" with and without adding detergent.

Materials

☐ A bowl of tap water.
☐ A dropper bottle of corn oil.
☐ Access to daylight from an outside window. A less complete view of oil in water is possible by viewing the slick in the reflection of overhead room lights.
☐ A dropper bottle of soapy water.

(TO) form oil films on water that reflect color by light interference. To understand this phenomenon as a function of film thickness.

THIN SLICK **O** **Cohesion / Adhesion ( )**

1. Fill a clean bowl half full of tap water. Set it where you can see daylight reflected on the surface, and let it calm.

2. Dip the point of a pin into a drop of corn oil, and touch it to the water while watching the reflected light. Repeat to form a series of 5 small oil slicks arranged in a circle.

a. White light bounces off these oil slicks to form colored *interference* patterns. What colors can you see?

b. These oil slicks make wonderful patterns that change over time. Draw some of the more interesting shapes.

3. Cite experimental evidence to support both conclusions stated below. (Repeat the experiment, as necessary.)

a. Corn oil gradually weakens the surface "skin" of water.

b. Reflection of color by light interference decreases as the thickness of these oil slicks increase.

 16

Answers / Notes

2. *If oil slicks fail to spread out widely from the pin point so you see no color, the surface of your water is contaminated with traces of soapy water or corn oil. Discard this water, thoroughly rinse the bowl, and try again.*

2a. All colors of the rainbow are present: red, orange, yellow, green, blue and violet.

2b. *Shapes like these are common, and they continue to change over time:*

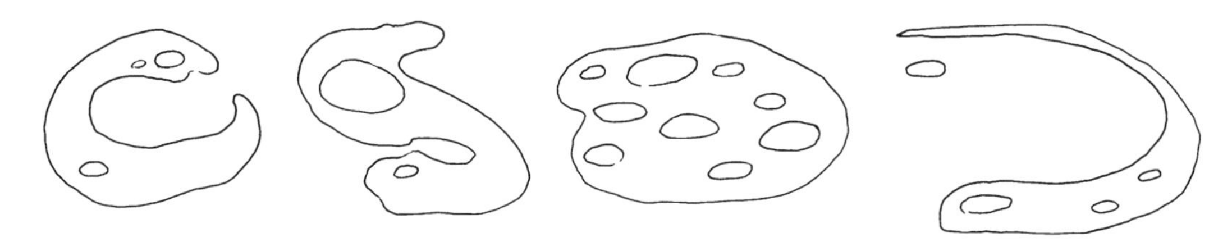

3a. When forming this series of 5 small oil slicks, each one spreads less rapidly and forms a smaller circle than the one before. This suggests that corn oil gradually weakens the surface tension of water. Adhesive forces between oil and water pull against a weakened water "skin" that is no longer as "rigid" a surface. This allows cohesive forces within the oil to increasingly dominate, confining the slicks to smaller and smaller circles.

3b. As these oil slicks get smaller, they become progressively less colorful, until there is no reflected surface interference at all. Assuming approximately equal volumes of oil are released from the pin into each new slick, their reduced diameters imply an associated increase in thickness.

Materials

- ☐ A clean bowl of tap water.
- ☐ Access to daylight, or full-spectrum electric light.
- ☐ A pin.
- ☐ A dropper bottle of corn oil.
- ☐ A piece of waxed paper to hold the drop of corn oil (optional). Or drip the corn oil directly on your desk.

(TO) associate reduced surface tension in water with the formation of long-lasting surface bubbles. To admire the varied color patterns that result when light reflects off the surface of a thin bubble film.

BUBBLES O **Cohesion / Adhesion ( )**

1. Fill a clean glass 1/3 full of fresh water. Blow air bubbles through this water with a clean straw.

a. What happens to these air bubbles when they reach the surface?

b. Add a dropperful of soapy water to the glass, and blow more bubbles through the water. Do they behave differently than before? Explain.

c. What is the relationship between surface tension and surface bubbles?

2. Cover a plastic lid with several droppers of soapy water.

a. Dip your straw to the bottom of the soapy water bottle, then blow a dome-shaped bubble on the wet surface of the lid. (Pull out the straw before your dome pops.)

b. Describe how the surface of your bubble changes in reflected light. Can you predict when the dome will break?

 17

Answers / Notes

1a. Air bubbles that form under water break immediately when reaching the surface. *(Occasionally a small surface bubble may linger against the side of the glass, but it doesn't last long.)*

1b. Yes. With soap in the water, air bubbles rise to the surface and accumulate as a surface foam. These bubbles remain for some time before popping.

1c. There is an inverse relationship. Surface bubbles do not easily form on water with high surface tension. Those that do pop quickly. If surface tension is weakened by the addition of soap, however, surface bubbles form easily and accumulate as foam.

Strong surface tension doesn't allow water to be stretched into a thin surface film for more than an instant before strong cohesive forces pull the water molecules back together into small compact drops. If you weaken these cohesive forces with soap, however, the thin film of water molecules will no longer coalesce. The bubble remains stable until its top surface, drained by gravity and thinned by evaporation, finally breaks.

2a. *Dipping the straw to the* bottom *of the bottle wets a maximum length of straw with bubble solution. The bubble dome may break prematurely if it touches any part of the straw or lid that is* dry, *or if it is evaporated rapidly by a breeze.*

2b. A fresh new bubble has a glassy appearance that reflects light. Soon a bull's-eye of circular color bands form around the top of the bubble dome, enclosing a clear, unreflective spot in the middle. These color bands gradually widen and move downward, while dull, opaque, cloud-like swirls develop within the enlarging clear spot. This gathering of dull swirls forecast that the bubble will soon break, several minutes after it was formed.

If you substitute bubble solution from activity 20 that contains glycerin, a similar but more dramatic surface evolution emerges: the bands of color grow larger and more intense, extending down to the very base of the dome; dramatic patterns of cloud-like opaqueness swirl inside a much larger clear area at the top of the dome before the bubble breaks.

Materials

☐ A glass or jar of water and a plastic straw.

☐ A dropper bottle of soapy water and a plastic lid. Or substitute the bubble solution from activity 20 and use the lid of its holding container. Both soapy water and bubble solution form stable bubble domes. The difference between these solutions is that glycerin seems to produce surface colors and patterns that are more intense.

(TO) graph how wave trains interfere constructively when they are in phase, and destructively as they shift out of phase.

INTERFERENCE

Cohesion / Adhesion ( )

1. Cut out the top rectangle from the Interference sheet. Graph how the *first* set of waves (upper + lower) add together to total one taller wave.

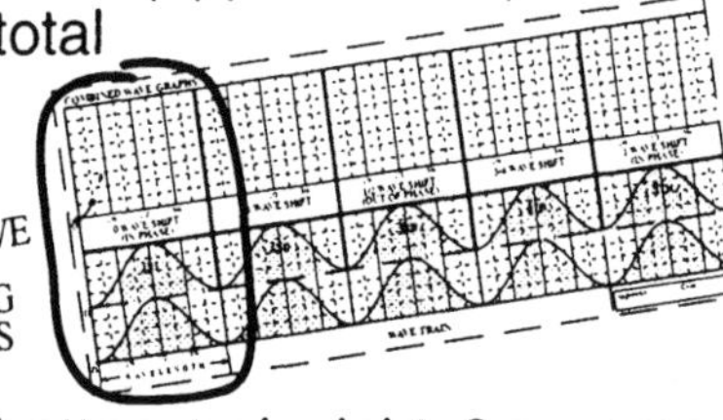

2. Cut off the bottom Wave Train on the dashed line. Shift it 2 squares *right* (1/4 wave length) and tape lightly. Graph how the *second* set of waves now combine.

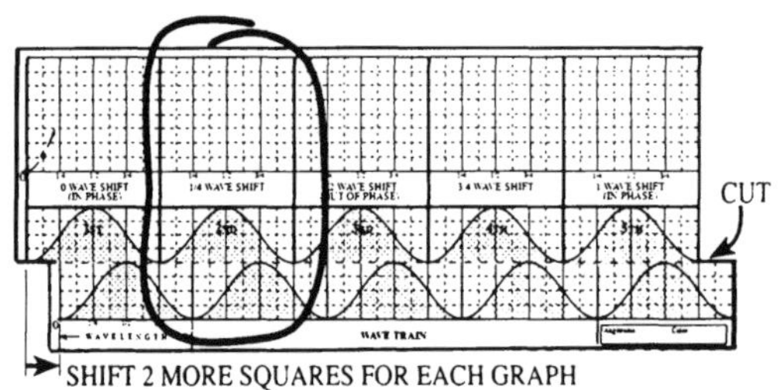

3. Shift the bottom train right, 2 squares at a time, to finish the last three graphs. (Save the bottom train for the next activity.)

4. Rephrase each statement below in terms of the graphs that you have drawn:
   a. Waves moving in phase interfere constructively.
   b. Waves moving out of phase interfere destructively.

5. *Coherent* (in phase) wave trains of blue light are shifted by these wavelengths. Complete the table.

WAVE SHIFT	0	1/4	1/2	3/4	$1\frac{1}{2}$	2	$3\frac{3}{4}$
BRIGHTNESS	*high*	*med*	*zero*				

 18

Answers / Notes

1. *Students should not shade their graphs. We shaded the area beneath both wave trains to make them visually easy to combine on the graph. This shading does not represent actual wave shapes, which oscillate above and below a midline like this:*

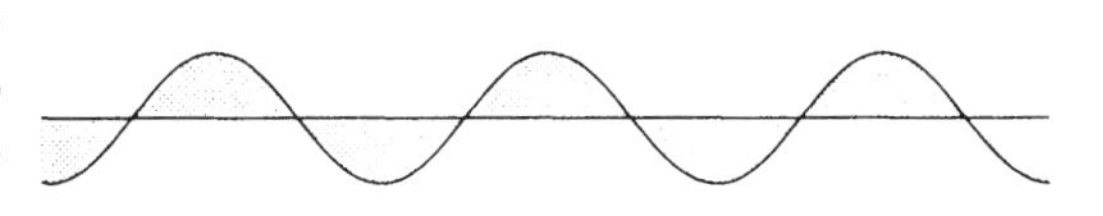

2. *Suggest that students reduce the tackiness of their tape before joining the graphs. They can do this by touching it to skin several times before applying it to the paper.*

1-3.

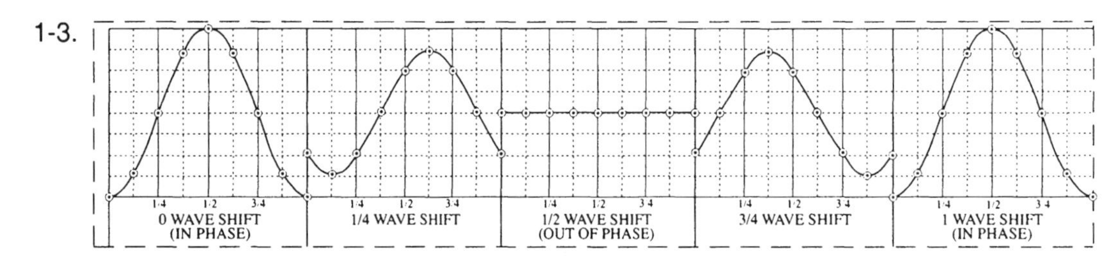

4a. Waves that move in step (matching trough to trough and peak to peak) combine in the tallest possible wave.

4b. As the wave trains begin to shift out of step, their combined amplitudes are no longer as tall. When they are completely out of phase (trough to peak and peak to trough), they cancel each other out, producing no wave at all.

5.

WAVE SHIFT	0	1/4	1/2	3/4	$1\frac{1}{2}$	2	$3\frac{3}{4}$
BRIGHTNESS	*high*	*med*	*zero*	med	zero	high	med

Materials

☐ The upper part of the supplementary sheet labeled "Interference." Photocopy this from the back of this module. (Save the lower part labeled "Thin Film" for the next activity.)
☐ Scissors.

(TO) model how thin oil films and thin soap films interfere with white light to produce a spectrum of color.

WAVE SHIFT

Cohesion / Adhesion ( )

1. Cut out the WaveTrain strip on the Thin Film sheet. This represents a light wave enlarged about 40,000 times, to the same scale as this Wavelength Ruler. ⟶

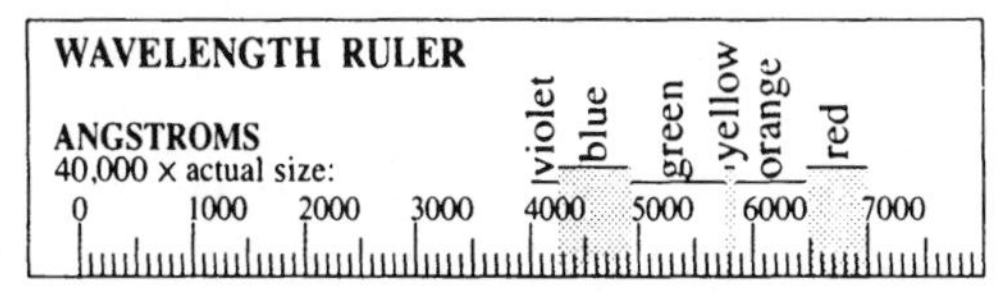

2. Measure the length of this light wave in Angstroms. Write this measurement, and the color we see at this wavelength, in the boxed space on the right.
3. Identify the wavelength and color of the wave train you graphed in the last activity. Fill in its box on the right.
4. Cut out the 2 Thin Film strips. Tape them end to end.
   a. This shows a cross section of thin film. Why does it look so thick?
   b. Six rays of white light strike this film. Tell what happens to ray "A."

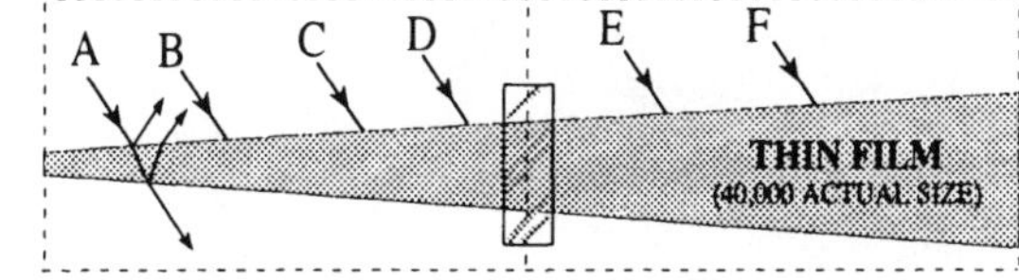

5. Use both wave trains like rulers. Which rays of white light...
   a. Interfere destructively to cancel blue? Yellow?
   b. Interfere constructively to reinforce blue? Yellow?

 19

Answers / Notes

2. One wavelength of this new wave train measures 5800 Angstroms on the ruler, corresponding to the middle of the yellow range. *(Students should write this length and "yellow" in the box at the lower right corner of the wave train.)*

3. One wavelength of the previous wave train measures 4670 Angstroms on the ruler, corresponding to the middle of the blue range. *(Again, students should write this wavelength and color in the box at the right.)*

4a. The film appears thick because it is drawn 40,000 times larger than actual size. *(The variable thickness of the film in our model simulates the thin-to-thick wall of bubble film created by downward drainage.)*

4b. Ray "A" reflects and refracts at both the tip and bottom surfaces of the thin film. The reflected parts of this ray are phase-shifted, because the bottom-reflected light has traveled farther than the top-reflected light.

5. *Students should measure the extra distance travelled (down and back) by the bottom-reflected light against both wavetrain "rulers": if this distance equals 1/2 wavelength (or a multiple thereof), that color is out of phase, and therefore canceled; if it equals a whole wavelength, that color is in phase, and therefore reinforced.*

5a. Ray A shifts 1/2 wavelength (1/4 down plus 1/4 up) on the "blue ruler" to cancel blue. (Other colors in the white light are still reflected to varying degrees.)

Ray B shifts 1/2 wavelength on the "yellow ruler" to cancel yellow. (Other colors in the white light are still reflected to varying degrees.)

Ray E shifts 1 1/2 wavelengths on the "blue ruler" to cancel blue.

Ray F shifts 1 1/2 wavelengths on the "yellow ruler" to cancel yellow.

5b. Ray C shifts 1 wavelength on the "blue ruler" to reinforce blue. (Blue is reflected from this part of the film more intensely than other colors.)

Ray D shifts 1 wavelength on the "yellow ruler" to reinforce yellow. (Yellow is reflected from this part of the film more intensely than other colors.)

Materials

☐ The supplementary sheet labeled "Thin Film." This is the bottom part of the page photocopied for the last activity.
☐ Scissors.
☐ Clear tape.
☐ The wave train saved from the last activity.

(TO) observe bands of color reflected in soap film. To interpret this phenomenon in terms of light interference and variable film thickness.

COLOR BANDS ○ Cohesion / Adhesion ( )

1. Write your name on the lid of a small tub. Pour in bubble solution to cover the bottom.

2. Hold a paper cup upside down, and dip its mouth in the bubble solution.

3. Tilt the cup up at an angle so that light from a window reflects in the soap film. Observe what happens in the film.

   a. Describe in words and drawings how the film's surface changes over time.

   b. As gravity pulls solution downward, bands of color appear. Explain why this happens.

4. Repeat the experiment. When the top color band has drained a third of the way down the film, hold the cup level.

   a. Watch reflected light while blowing gently on the film with a straw.

   b. Describe what you see.

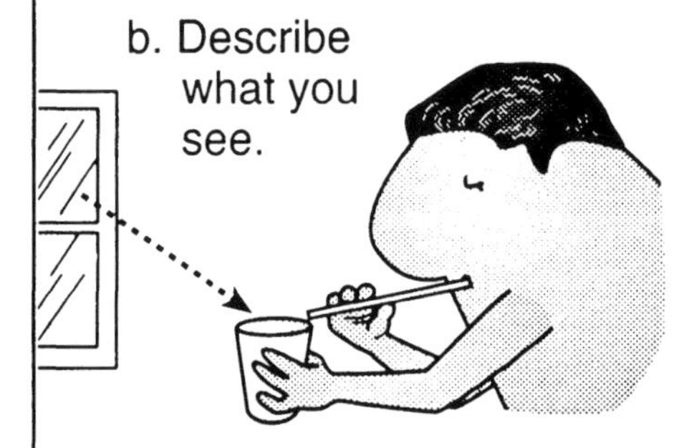

5. Cover your liquid and save it for all remaining activities.

 20

Answers / Notes

3a. Bands of color form at the top of the ring and gradually migrate downward. A reflective surface with less color forms above these bands, and a very thin, unreflective surface forms above that. These reflective and transparent surfaces mix in complex swirls as they are agitated by air currents. The reflective area increases in color and complexity of pattern as more film drains downward. When only about 1/6 of the film still reflects color, it is about to break.

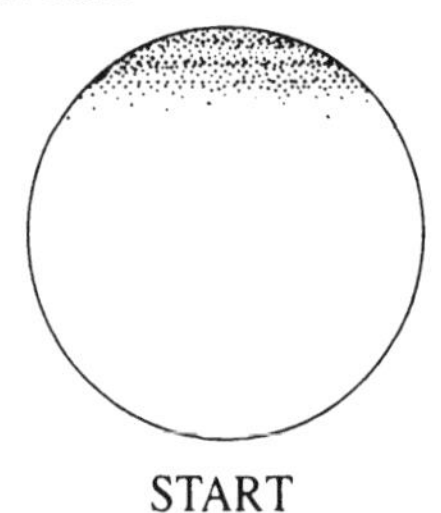

START

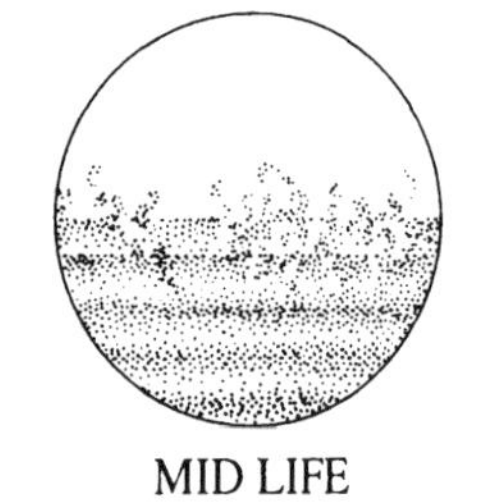

MID LIFE

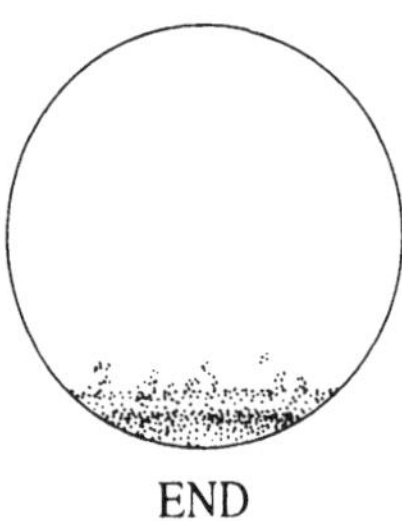

END

3b. Gravity drains solution down the film's surface, creating a thinner layer at the top, a thicker layer toward the bottom. White light, reflecting off both the inner and outer surfaces of the film, results in repeating colored bands of interference. The blue component of the light, for example, phase-shifts constructively and destructively between these two reflections, producing repeating bands of blue depending on the thickness of the film.

4b. Brilliant oranges, yellows, greens, blues and purples swirl in rich, fascinating patterns of color.

Materials

☐ A plastic tub (margarine or deli container) with tight-fitting lid.
☐ Masking tape (for name labels).
☐ A styrofoam (or paper) cup.
☐ Bubble solution prepared *in advance.* (It improves with age.) To a clean, plastic gallon (4 L) milk jug, add:
✔ ½ cup (120 mL) dishwashing liquid. Joy* or Dawn* brands work best. Never use a low-suds detergent.
✔ ¼ cup (60 mL) glycerine, available in most pharmacies.
✔ Almost 1 gallon (4L) tap water. If your water is hard, substitute distilled or deionized (soft) water. Gently pour it down an inner wall of the jug (to avoid suds) until nearly full. Cap tightly. Turn the jug up and down to mix. (More concentrated solutions increase bubble strength only marginally.)
☐ A plastic drinking straw.
☐ Access to daylight or a full-spectrum bulb.

(TO) observe how cohesive forces in soap film contract it into a shape that minimizes total surface area.

LOOP THE LOOP O Cohesion / Adhesion (   )

1. Tie a snug loop of thread around a battery. Trim one end short; leave the other end long, then slide it off.

2. Pin your loop into an Area Grid backed by cardboard to form each figure below.
   a. Count the squares in each regular polygon.
   b. Which shape encloses the maximum area?

2-PIN LINE ⁄	*(no area)*
EQUILATERAL TRIANGLE △	
SQUARE ▢	
REGULAR PENTAGON ⬠	
PERFECT CIRCLE ◯ *(trace base of battery)*	

3. Cover the mouth of a cup with bubble solution, as before. Set the cup upright on your table.

4. Wet your thread loop in the solution. Gently lay it in the middle of the film surface.

5. Break the film *inside* the loop with dry paper or a dry finger. What happens?

6. Explain in terms of ***cohesion***...
   a. ...what the loop's shape tells about the area of film that remains.
   b. ...why the film moves.

 21

Answers / Notes

2a. Squares that fall inside each shape by *more* than half should be counted as "1." Those that are *less* than halfway inside should be counted as "0."

2-PIN LINE ⁄	*(no area)*
EQUILATERAL TRIANGLE △	*114 squares*
SQUARE ▢	*144 squares*
REGULAR PENTAGON ⬠	*170 squares*
PERFECT CIRCLE ◯	*188 squares*

2b. The circle (a regular polygon with an infinite number of equal sides) encompasses the greatest area.

5. When the soap film inside the loop is broken, the thread is drawn into a perfect circle. Further, the bubble film creeps down the inside of the cup toward the bottom.

6a. Cohesive forces in the unbroken film outside the loop contract equally in all directions to achieve a *minimum* surface area. This pulls the empty loop into a perfect circle that surrounds the *maximum* possible opening.

6b. The same forces draw the remaining film toward the smallest available opening, which is at the bottom of the cup. *(While these cohesive forces are weak compared to those of water, they are strong enough to minimize surface area. Students will explore more of this fascinating geometry in the next two activities.)*

Materials

- ☐ Thread.
- ☐ A size-D battery, dead or alive.
- ☐ Scissors.
- ☐ An Area Grid. This was previously photocopied along with the Water Molecules cutout in Activity 5.
- ☐ A piece of corrugated cardboard large enough to contain the Area Grid.
- ☐ Straight pins.
- ☐ A cup (paper cups work as well as or styrofoam in this activity).
- ☐ The tub with bubble solution from Activity 20.
- ☐ A piece of paper towel, tissue or other dry paper to break the bubble film (a dry finger will also work).

(TO) recognize that cohesive forces always contract soap film into shapes that minimize surface area, in both two and three dimensions.

MINIMUM SURFACES ○ Cohesion / Adhesion ( )

1. Dip the top of 2 batteries in bubble solution. Blow small bubbles of equal size on the top each battery through a straw.

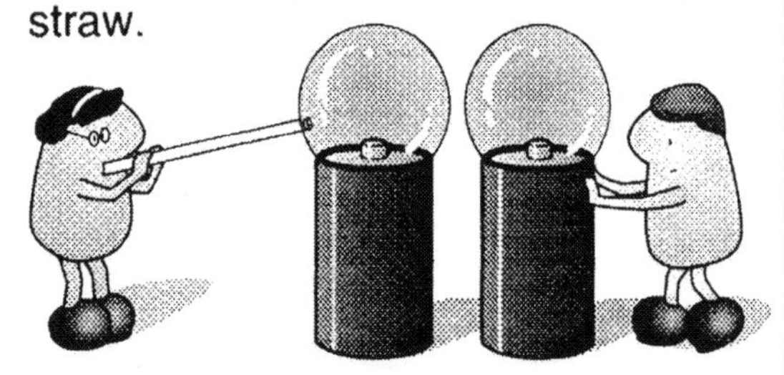

a. Why are these bubbles shaped like spheres, and not cubes?
b. Touch the bubbles together. Draw a side view showing how these films combine to reduce surface area.
c. Do bubbles of *unequal* size combine to form a *flat* common wall? Draw your findings.

2. Cut the bottom off a tapered cup. Wet it inside with bubble solution.

a. Dip the wide end in the solution to form a film over its mouth.
b. What happens? Why?

 22

Answers / Notes

1a. The bubbles take the shape of spheres, not cubes, because this shape encloses a maximum volume of air with a minimum area of soap film. *(This corresponds to a 2-dimensional circular thread loop enclosing more surface area than a square loop.)*

1b. Both bubbles reduce their surface areas by forming a common flat wall between them. Sometimes this inner wall breaks, reducing the surface area of the film even more.

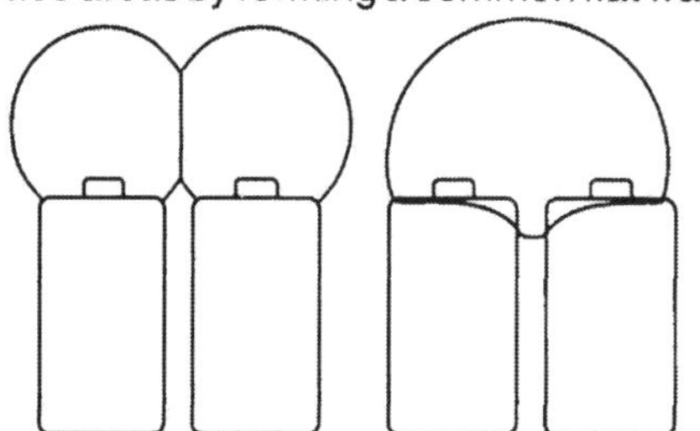

1c. No. The common wall curves into the larger bubble.

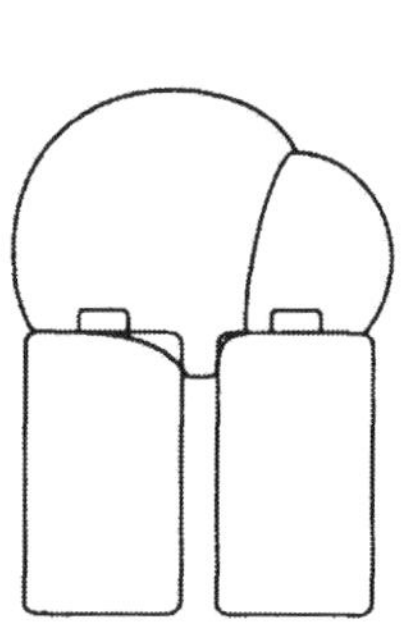

2b. When the cup is dipped into the solution, a soap film covers its mouth. Cohesion, contracting this surface area to a minimum, slowly pulls the film from the wider mouth of the cup to its narrower base. This works even against gravity.

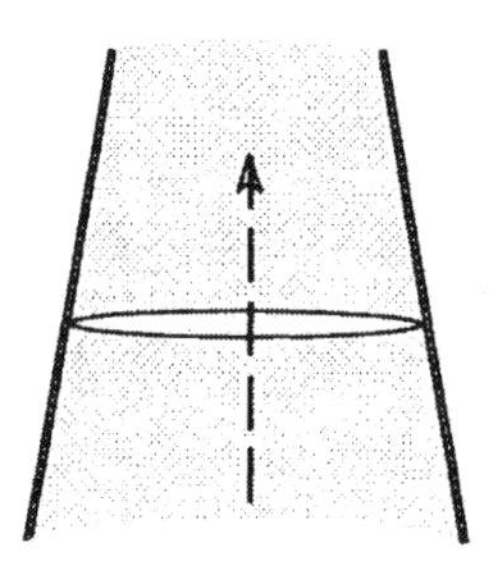

Materials

- ☐ Two size-D batteries.
- ☐ The bubble solution in a deli container.
- ☐ A plastic straw.
- ☐ A tapered cup made of waxed paper or styrofoam.
- ☐ Scissors.

(TO) furthur explore the geometry of soap films.

BUBBLE ARCHITECTURE ○ Cohesion / Adhesion (   )

1. Carefully cut an even ring from a foam or paper cup about 3 cm ($1^1/_4$ inch) from the rim.

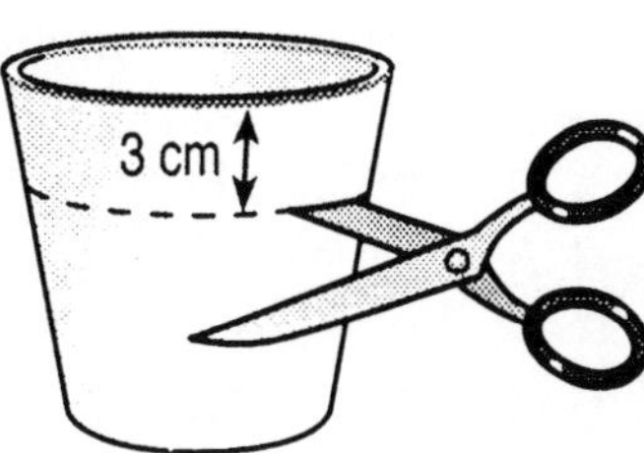

SMOOTH CUT EDGE

a. Dip the smaller end of this ring and the rim of a second cup in bubble solution. Touch the two film windows together, then pull them a little apart.

b. Draw what you see.

c. Break the center circle of film with a dry finger. Draw the film shape that results.

2. Use a straw to blow a layer of small connected bubbles on the wet lid of your tub.

a. Find these shapes between the bubbles: **triangle**; **square**; **pentagon**; **hexagon**. Describe what causes each shape.

b. State a general rule for making a regular polygon with **n** sides.

 23

Answers / Notes

1b. As the rings are moved apart, the joined bubble films create two curved funnels that narrow to a common central window.

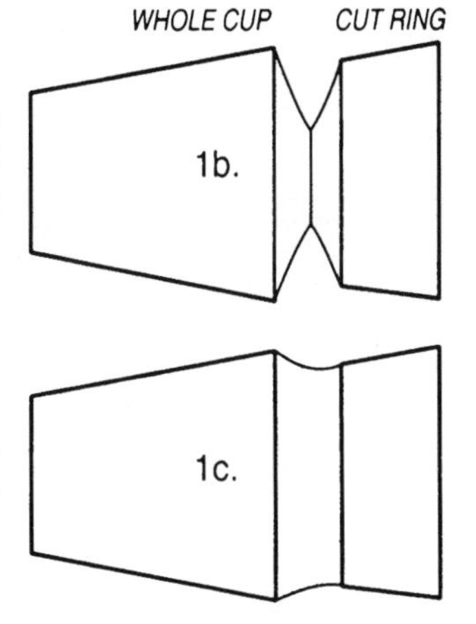

1c. When you break the shared film "window," the film contracts into an inward-curved tube connecting the rings.

2a. The number of bubbles one layer deep that surround a central bubble determine its geometry:

3 exterior bubbles of equal size push the central bubble on all sides to form an equilateral *triangle.*

4 exterior bubbles of equal size reshape the central bubble into a *square.*

5 exterior bubbles make a *pentagon.*

6 exterior bubbles produce a *hexagon.*

2b. General rule: ***n*** exterior bubbles of equal size push against a central bubble on all sides to form a regular polygon with ***n*** sides.

Extension

Build wire frames of different shapes and submerge them in bubble solution. Sketch the geometry of minimum surface shapes. Here are four examples:

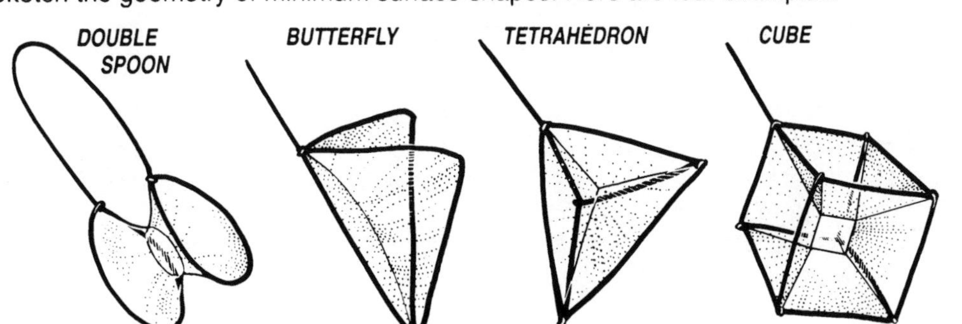

Materials

☐ Two cups (paper cups work as well as or styrofoam).
☐ Scissors.
☐ The tub of bubble solution from previous activities.
☐ A drinking straw.

(TO) explore variables that influence bubble size.

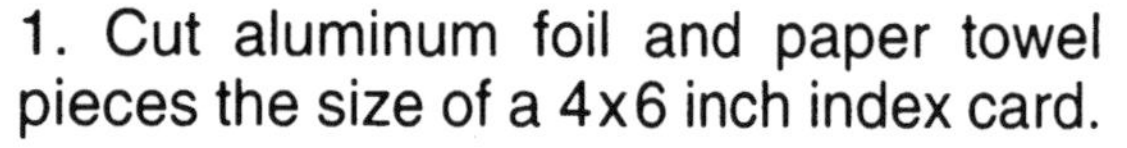

GIANT DOMES ○ Cohesion / Adhesion ( )

1. Cut aluminum foil and paper towel pieces the size of a 4x6 inch index card.
2. Fold each in half the long way. Join with masking tape, front and back, with the folded edges to the outside.
3. Roll the center tape around a narrow test tube. Fix it with more tape, leaving both ends free.

4. Slide your cylinder off the test tube, and soak the paper end in bubble solution.
5. Pour a puddle of bubble solution in the middle of your table. Blow into this puddle through the foil to make *giant* domes.

6. Measure the diameter of the ring left by the largest popped dome. Go for a record!
7. Can you blow giant domes through a straw? If you wrap a strip of soapy towel around the end, does this make a difference?
8. What variables influence dome size? Write a report.

 24

Answers / Notes

3. *The finished cylinder looks like this:*

(ACTUAL SIZE)

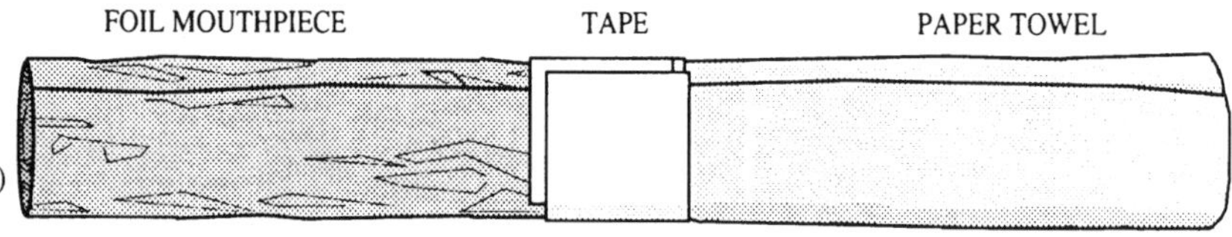

6. *We have achieved bubble domes over 1/2 meter in diameter using the bubble solution recipe in activity 20.*

7. Bubble domes about half this diameter (25 cm) can be blown through a simple straw. Adding a strip of soapy paper towel to the tip extends the size limit to about 40 cm.

8. These variables influence dome size:
a. Quantity of bubble solution: Larger bubble domes use more solution. This is delivered to the dome by larger diameter blowing devices, and/or saturated paper towel.
b. Time: Gravity drains solution downward, thinning the top of the dome until it breaks. The faster you blow the domes, therefore, using a large diameter tube, the larger it can grow before thinning to its breaking point.
c. Technique: Blowing into the bubble dome from the top downward (rather than horizontally from the side) introduces fresh soap solution into the thinner upper dome, thereby slowing the thinning process.
d. Formula of bubble solution: Certain brands of liquid soap produce longer lasting bubbles than others. Dilution has some effect, as well. *(Other additives that strengthen soap film, beside glycerin, include unflavored gelatin and Karo Syrup. Could water be left out of the formula entirely? Inventing a stronger bubble solution might develop into a science project.)*

Materials

- ☐ A 4x6 inch index card.
- ☐ Aluminum foil.
- ☐ Paper towels.
- ☐ Scissors.
- ☐ Masking tape.
- ☐ A meter stick.
- ☐ A medium-sized test tube with a diameter of about 1.5 cm. Or substitute other cylindrical objects (wood dowel, marking pen, etc.) of comparable diameter.
- ☐ The bubble solution in a shallow deli tub.
- ☐ A standard-sized plastic straw, about 1/4 inch in diameter.
- ☐ A bottle of vinegar (optional). Use this to cut residual soap film on desk tops and facilitate cleanup.

REPRODUCIBLE STUDENT TASK CARDS

☞ As you duplicate and distribute these task cards, **please observe our copyright restrictions** at the front of this book. Our basic rule is: **One book, one teacher.**

☞ TOPS is a small, not-for-profit educational corporation, dedicated to making great science accessible to students everywhere. Our only income is from the sale of these inexpensive modules. If you would like to help spread the word that TOPS is tops, please request multiple copies of our **free TOPS Ideas catalog** to pass on to other educators or student teachers. These offer a variety of sample lessons, plus an order form for your colleagues to purchase their own TOPS modules. Thanks!

Task Cards Options

Here are 3 management options to consider before you photocopy:

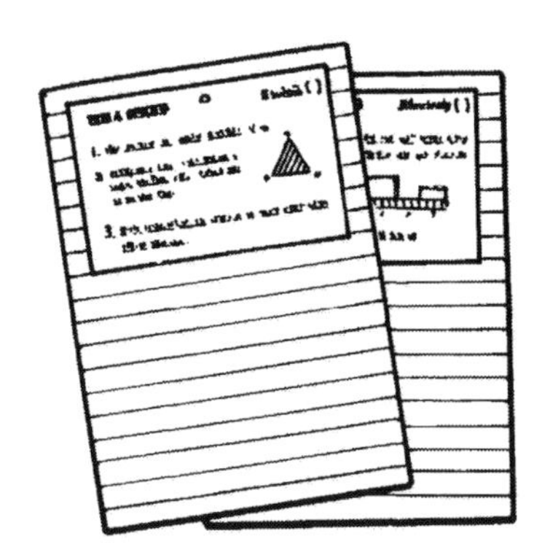

1. Consumable Worksheets: Copy 1 complete set of task card pages. Cut out each card and fix it to a separate sheet of boldly lined paper. Duplicate a class set of each worksheet master you have made, 1 per student. Direct students to follow the task card instructions at the top of each page, then respond to questions in the lined space underneath.

2. Nonconsumable Reference Booklets: Copy and collate the 2-up task card pages in sequence. Make perhaps half as many sets as the students who will use them. Staple each set in the upper left corner, both front and back to prevent the outside pages from working loose. Tell students that these task card booklets are for reference only. They should use them as they would any textbook, responding to questions on their own papers, returning them unmarked and in good shape at the end of the module.

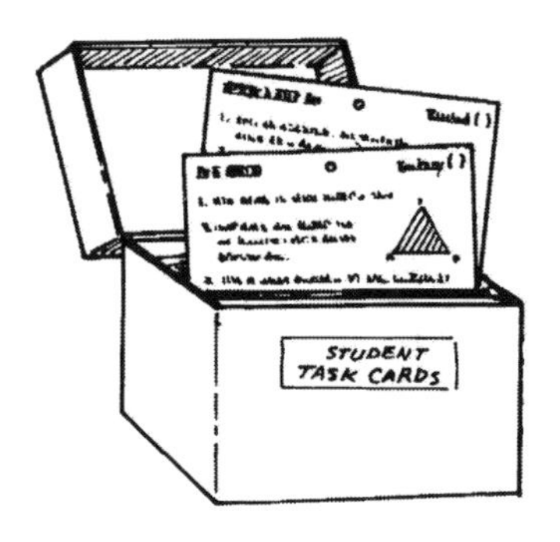

3. Nonconsumable Task Cards: Copy several sets of task card pages. Laminate them, if you wish, for extra durability, then cut out each card to display in your room. You might pin cards to bulletin boards; or punch out the holes and hang them from wall hooks (you can fashion hooks from paper clips and tape these to the wall); or fix cards to cereal boxes with paper fasteners, 4 to a box; or keep cards on designated reference tables. The important thing is to provide enough task card reference points about your classroom to avoid a jam of too many students at any one location. Two or 3 task card sets should accommodate everyone, since different students will use different cards at different times.

COHESION

O **Cohesion / Adhesion ( )**

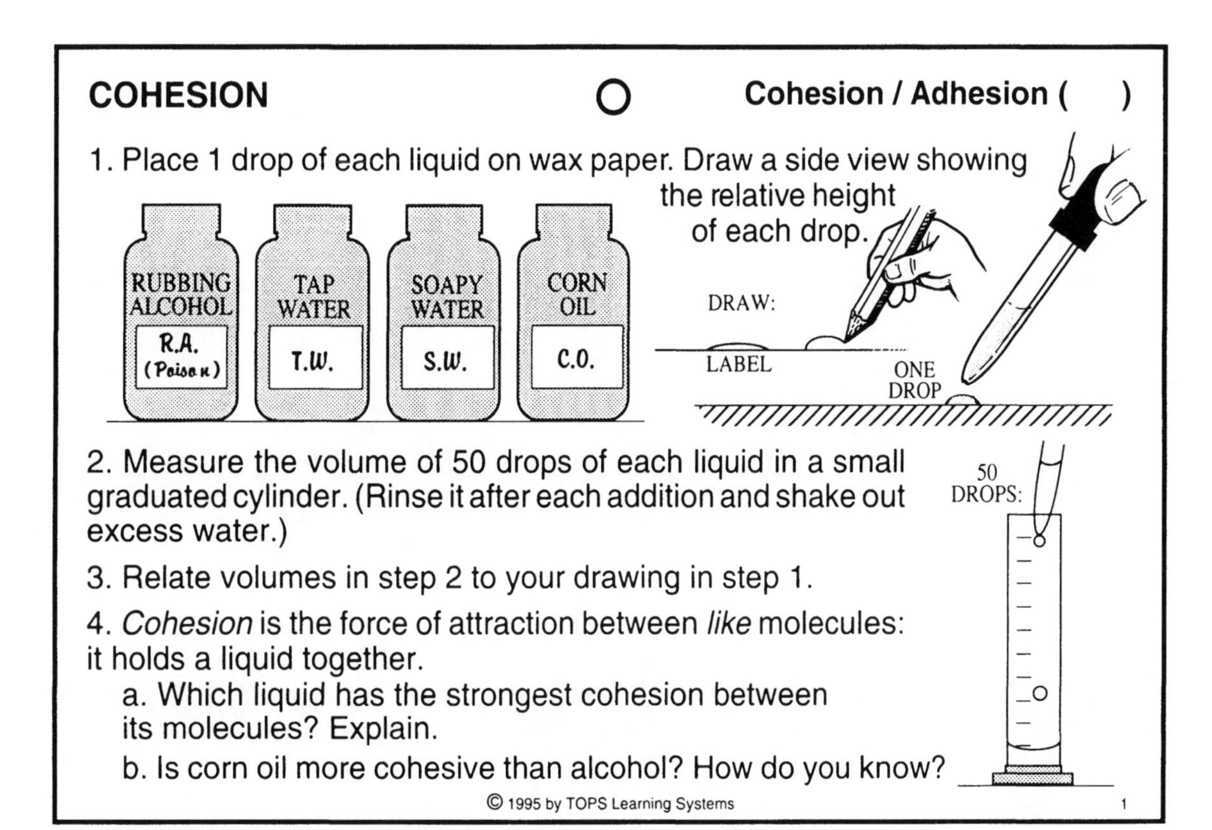

1. Place 1 drop of each liquid on wax paper. Draw a side view showing the relative height of each drop.

2. Measure the volume of 50 drops of each liquid in a small graduated cylinder. (Rinse it after each addition and shake out excess water.)

3. Relate volumes in step 2 to your drawing in step 1.

4. *Cohesion* is the force of attraction between *like* molecules: it holds a liquid together.
   a. Which liquid has the strongest cohesion between its molecules? Explain.
   b. Is corn oil more cohesive than alcohol? How do you know?

 1

HEAP O' WATER

O **Cohesion / Adhesion ( )**

1. Calculate how many drops of tap water add up to 1 mL. Set up a proportion using numbers from the last activity.

2. Try to heap 1 mL of tap water on top of a penny without spilling any over its edge. Explain your results in terms of cohesion.

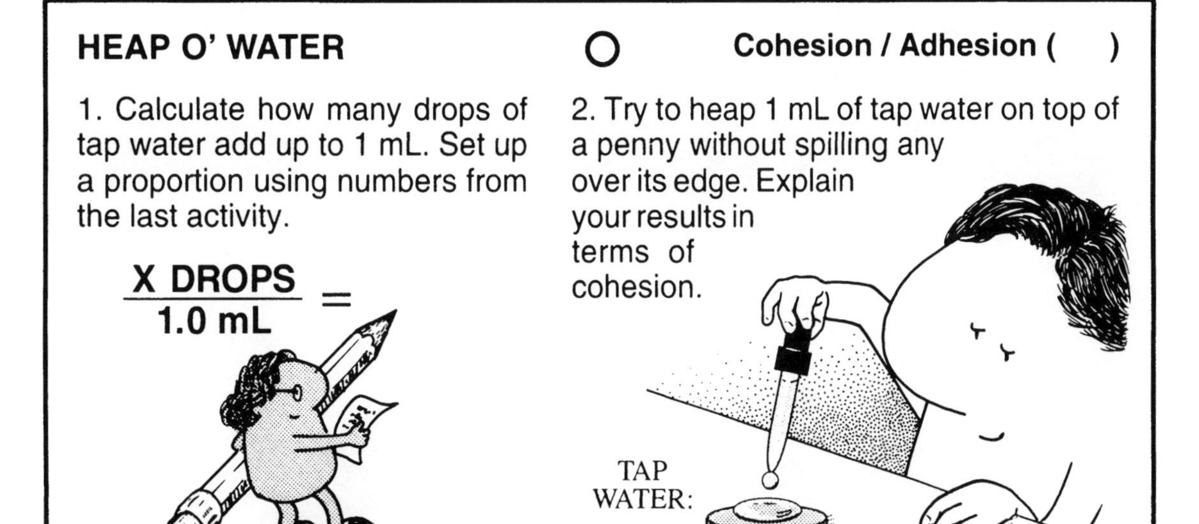

3. Dry the penny. Add 1 drop *less* than a full mL of tap water to the penny without spilling.
   a. Now add 1 drop more of soapy water. What happens?
   b. How does soap affect the cohesion of water?

 2

ADHESION

○ Cohesion / Adhesion ( )

1. Tear off a square-shaped piece of wax paper as wide as the roll, and fold to one-quarter size. Trim, tape and label the torn edge as shown:

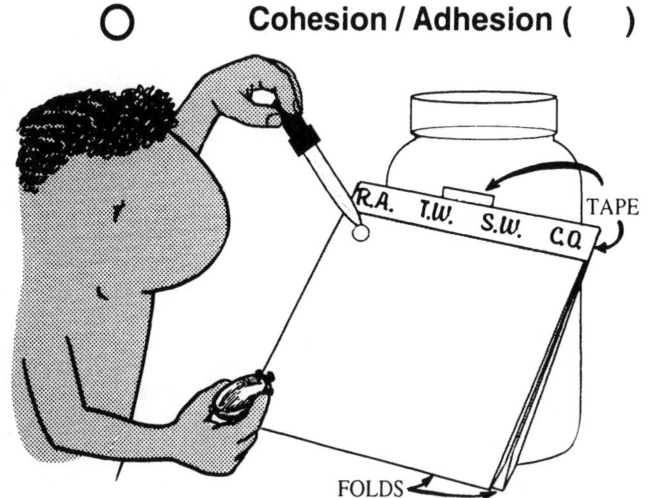

2. Tape it against a jar to form a steep ramp. Time how many seconds it takes for one drop of each liquid to move from the top of this ramp to the bottom.

3. *Adhesion* is the force of attraction between *unlike* molecules that allows the surface of one substance to stick to another. Which liquid adheres to wax paper with the greatest force? Least force? Explain.

4. Draw the trails left by rubbing alcohol and corn oil on wax paper. Relate these patterns to adhesion.

CREEPY CRAWLIES

Cohesion / Adhesion ( )

1. Form 3 separate puddles of corn oil on wax paper, with 2 drops in each puddle.

2. Now add 1 drop of each liquid to the *edge* of a puddle. Aim well, so half of each drop touches the wax paper while the other half touches the oil.

CORN OIL (2 drops) — TAP WATER (1 drop) — CORN OIL (2 drops) — RUBBING ALCOHOL (1 drop) — WAX PAPER

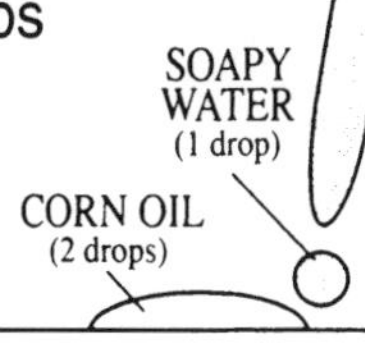

3. Here is a top view of *initial* and *final* resting places for tap water and oil.

INITIAL: CORN OIL, TAP WATER → FINAL: CORN OIL, TAP WATER

a. Make similar labeled drawings for all 3 puddles.
b. Explain how forces of adhesion and cohesion act on each liquid combination to rearrange it.

4. Make a 4-drop oil puddle with a 1-drop tap water "eye" in the center.

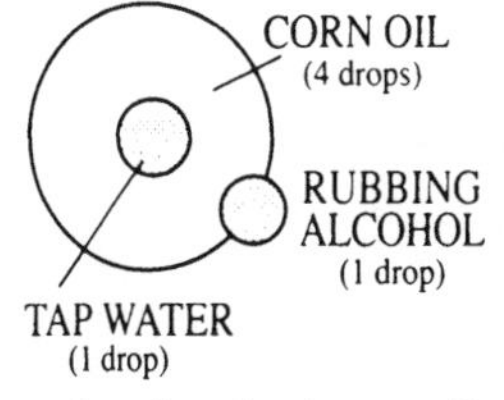

a. Add 1 drop of alcohol so it touches *only* the edge of the oil puddle but *not* the middle eye.
b. Describe the delayed reaction in terms of cohesion and adhesion.

H_2O O Cohesion / Adhesion ( )

1. Trim the Water Molecules cutout around its outer solid line. Cut along the thin dashed lines as follows:

   a. Fold along center line A. Cut along the 7 dashed lines to each *molecule*.

   b. Repeat along lines B and C.

   c. Cut in from edges D and E.

2. Stretch these molecules into a honeycomb: twist the left one up, the right one down, and pull gently apart.

   a. Fold the tabs over the rim of a paper plate and tape in place.

   b. Label your model like this:

3. One oxygen atom and two hydrogen atoms form a molecule of water joined by two covalent bonds. Water molecules have electron-rich areas and electron-poor areas that mutually attract to form hydrogen bonds.

   a. Copy the above description. Use your model to point out each underlined feature to a friend.

   b. Which bonds are stronger (closer together)? Which bonds are weaker (farther apart)?

 5

ZIPPERS O Cohesion / Adhesion ( )

1. Untape your Water Molecules model from the paper plate, and pull gently on the tabs. How does this model explain the source of cohesion in real water?

2. Fill a clean glass with tap water and sprinkle pepper on its surface.

   a. What happens when you drip 1 drop of soapy water down the inner wall of the glass?

   b. What happens when you stretch your Water Molecules model, then release one side?

   c. Explain how 2b models 2a.

3. Rinse your glass and set it in a bowl. Fill it higher than the brim with fresh tap water.

   a. Scrape a *tiny* speck off a bar of soap with a pin. Drop it on the water, and observe carefully.

   b. Where does this soap speck get its energy?

4. Add a few crumbs of solid camphor (the smaller the better) to a clean glass of tap water. Compare and contrast camphor specks with soap specks.

 6

SURFACE TENSION  Cohesion / Adhesion ()

1. Fill a bowl half full of tap water. Float a toothpick in the bowl.
2. Stick 2 pin "arms" in a bit of styrofoam as shown, to cradle a third pin sideways.
3. Gently lower the cradled pin onto the water beside the toothpick. *Surface tension* (a skin-like property of water) prevents the pin from sinking.
   a. Prove that the pin doesn't float like a toothpick.
   b. With your Water Molecules model and a pencil, model how real water supports the pin.
4. Support the pin with surface tension as before, next to the floating toothpick. Sprinkle pepper over the water.
   a. A pattern in the pepper tells you that the pin rests <u>on</u> water, while the toothpick floats <u>in</u> water. Explain. Illustrate your answer with a drawing.
   b. What will happen if you squeeze a dropper full of soapy water into the bowl? Make a reasoned prediction, then test it.

GROOVY  Cohesion / Adhesion ()

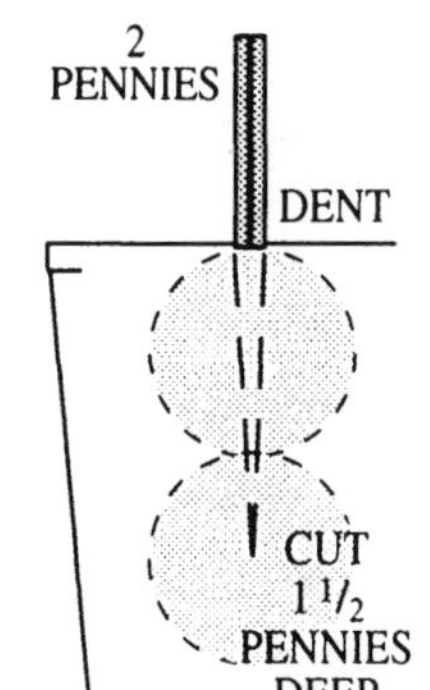

1. Press a dent into the lip of a styrofoam cup with the edge of 2 pennies. Cut a narrow V-groove into the cup, as wide as this dent at the lip, and about 1.5 penny diameters deep.
2. Tint a large jar of water with a drop of blue food coloring. Stand your V-cup in a bowl and gently fill it with blue water until it spills out the groove.
   a. How high can water rise in the cup? Describe how it spills down the side.
   b. Explain each observation in terms of surface tension, adhesion, and cohesion.
3. Spiral a blue stream around the white cup. Explain how each tool uses adhesion and cohesion to advantage:
   a. Use a *straw* to guide the stream in a spiral half-way around the cup.

   b. Rechannel the stream with a *wet string.* Spiral the stream in a full circle.

CAPILLARY ACTION  Cohesion / Adhesion (   )

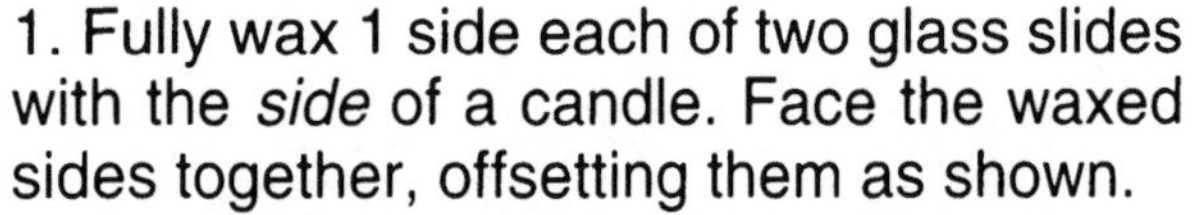

1. Fully wax 1 side each of two glass slides with the *side* of a candle. Face the waxed sides together, offsetting them as shown.

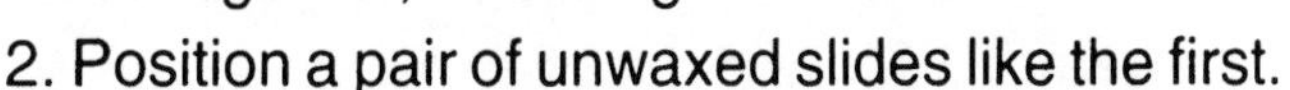

2. Position a pair of unwaxed slides like the first.
3. Place 1 drop of tap water on the exposed "lip" of each pair. Does water adhere with greater force to glass or to wax? Explain.
4. Move your head to see room light reflected on the unwaxed pair. Keeping this light in view, move the top slide over to just touch the water drop: draw and describe what you see.
   a. Water moves between the slides by a process called *capillary action.* Explain how adhesion and cohesion drive capillary action.
   b. Pick up the top slide. Why does the bottom slide stick to it?
5. Repeat step 4 for the waxed slides. Can you observe capillary action? Why?
6. Save both pairs of slides for the next activity.

 9

WALL 'O WATER  Cohesion / Adhesion (   )

1. Get both pairs of slides from the last activity.
2. Sandwich a paper clip between the unwaxed slides as shown, so it rests in an upper corner while you hold the opposite upper corner.
3. Dip the bottom edge into a bowl of water. Make a drawing to show how high water rises up between the slides.
4. Why does the water rise to form a "wall?" Why does this wall curve upward?

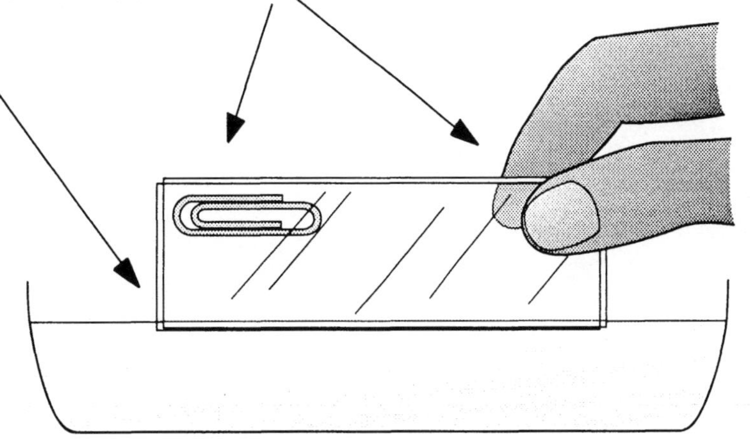

5. Repeat this experiment with your pair of waxed slides. Do you observe capillary action? Why?

 10

WATER LINES

O **Cohesion / Adhesion ( )**

1. Wrap 2 rubber bands vertically around a small, clean jar, so they run flat and parallel across the top.
2. Get a clean glass eyedropper tube with the rubber bulb removed. Stick it between the rubber bands, tip pointed up.
3. Set this jar in a bowl, and fill it brimful with tap water. Gently snap the tip of the tube with your finger to dislodge any water adhering inside the tip.
4. Copy a large side view of the jar and tube like this. Accurately draw how all curving surfaces of water meet the glass.

5. Note where different shapes in the water's surface are influenced by cohesion, adhesion, surface tension and capillary action.
6. How do you think your drawing would change if the glass tube were thinner?

CAPILLARY PATHWAYS

O **Cohesion / Adhesion ( )**

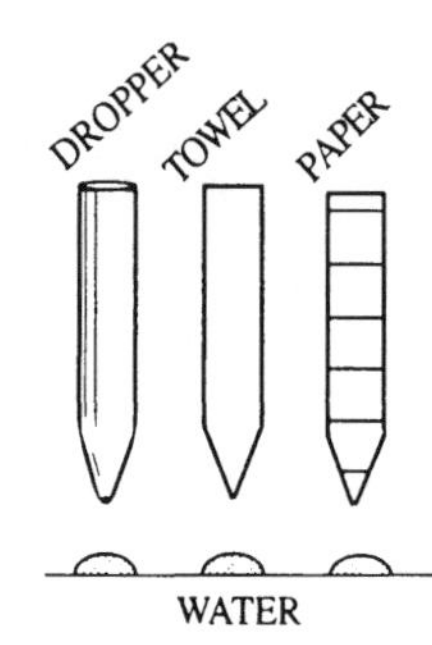

1. Get a clean glass eye dropper with the rubber bulb removed. Cut a strip of soft paper towel and a strip of notebook paper into a similar shape.
2. Put 3 separate drops of tap water on wax paper. Touch the point of each "dropper" to one of these drops.
   a. Which droppers have capillary action? Which do not?
   b. Examine the paper droppers with a hand lens. Propose a theory to explain why only one has capillary action.
3. Put water in a bowl, 1 cm deep, and add a drop of blue food coloring.
   a. Pinch a paper towel in the middle. Squeeze the rest into a narrow cone, and stand it in the water.
   b. What happens?
4. Wait at least 12 hours, then examine your cone again. How has it changed?
   a. Propose a theory to explain your observations.
   b. Explain how this models *transpiration,* how water moves through trees?

CHROMATOGRAPHY  Cohesion / Adhesion ()

1. Loop a rubber band "clothesline" across the mouth of a small dry jar.

2. Cut 2 strips, about 1 cm wide, from the white margin of a newspaper. Fold these as shown so they hang on your line and just touch the bottom of the jar.

3. Use washable markers to draw horizontal lines 1 cm above the bottom of each strip, one Black and one Brown. Label the tops with the colors you applied.

4. Hang these strips so they don't touch each other or the jar's sides. With an eyedropper, add clear tap water to wet only the *uncolored* "foot" of each strip.

5. Set up additional jars with Red, Orange, Yellow, Green, Blue and Purple.

6. Remove the strips before the colors reach the top of the jar. Tape them to your assignment sheet with writing space between them.
   a. Label all the colors you see in each strip.
   b. Which markers seem to contain a *mixture* of dyes? *Pure* dyes?

MORE CHROMATOGRAPHY  Cohesion / Adhesion ( )

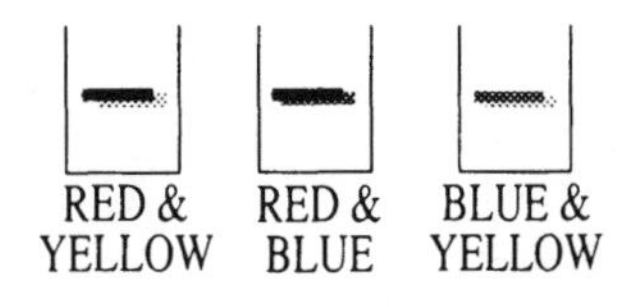

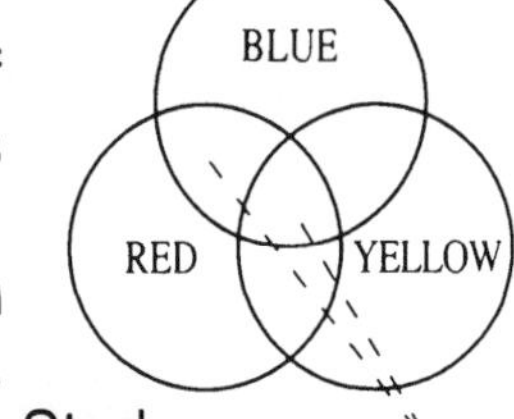

1. Superimpose these colored lines, one on top of the other, on strips of newspaper. Separate these pairs by chromatography as before.

2. Wait at least 10 minutes for the colors to separate, then tape the strips to your assignment sheet. Label all colors.
   a. Red, yellow and blue combine to create other colors. Study these and other chromatograms (color strips) to complete this color chart.
   b. Which dye molecules (red, blue or yellow) move slowest? Explain.

3. Cut into a piece of scratch paper about 1/5 of its length. Overlap and tape to form a scoop.

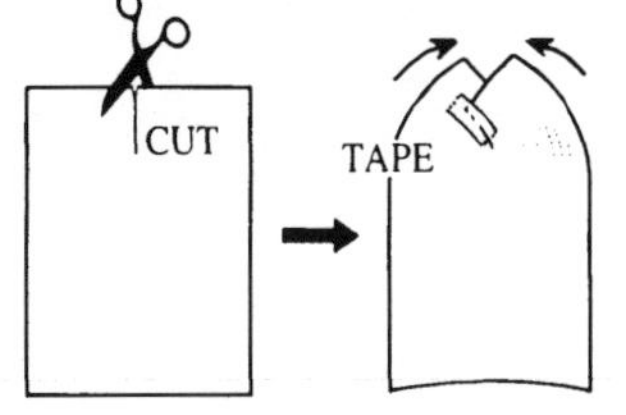

   a. Collect a pinch each of salt and pepper in the back of the scoop. Hold it at an angle and tap the end with your pencil to move these grains down its length.
   b. How do salt and pepper model the dye molecules moving through a chromatogram by capillary action?
   c. Why do some color combinations separate better than others?

SALT & PEPPER

THICK SLICK

Cohesion / Adhesion (   )

1. Put a bowl of tap water near a window where you can see daylight reflected on the water.

2. Let the water calm, then place 1 drop of corn oil on the surface. Watch the oil spread over the water in the reflected light.

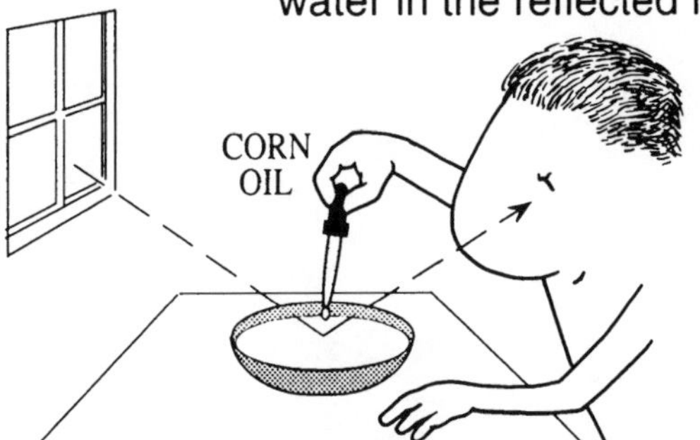

a. Draw pictures of the oil slick. How does it change over time?

b. Which is stronger, the cohesion of the corn oil, or its adhesion to the surface "skin" of water? Explain.

3. Add 1 drop of soapy water to the slick while watching it in reflected light.

a. What happened to the oil slick?

b. Did the cohesion of corn oil just get stronger, or the surface "skin" of the water just get weaker? Explain.

c. Cleanup crews often spray soapy detergents on oil spills. Explain how this makes their job easier.

THIN SLICK

Cohesion / Adhesion (   )

1. Fill a clean bowl half full of tap water. Set it where you can see daylight reflected on the surface, and let it calm.

2. Dip the point of a pin into a drop of corn oil, and touch it to the water while watching the reflected light. Repeat to form a series of 5 small oil slicks arranged in a circle.

a. White light bounces off these oil slicks to form colored *interference* patterns. What colors can you see?

b. These oil slicks make wonderful patterns that change over time. Draw some of the more interesting shapes.

3. Cite experimental evidence to support both conclusions stated below. (Repeat the experiment, as necessary.)

a. Corn oil gradually weakens the surface "skin" of water.

b. Reflection of color by light interference decreases as the thickness of these oil slicks increase.

BUBBLES

Cohesion / Adhesion (   )

1. Fill a clean glass 1/3 full of fresh water. Blow air bubbles through this water with a clean straw.

a. What happens to these air bubbles when they reach the surface?

b. Add a dropperful of soapy water to the glass, and blow more bubbles through the water. Do they behave differently than before? Explain.

c. What is the relationship between surface tension and surface bubbles?

2. Cover a plastic lid with several droppers of soapy water.

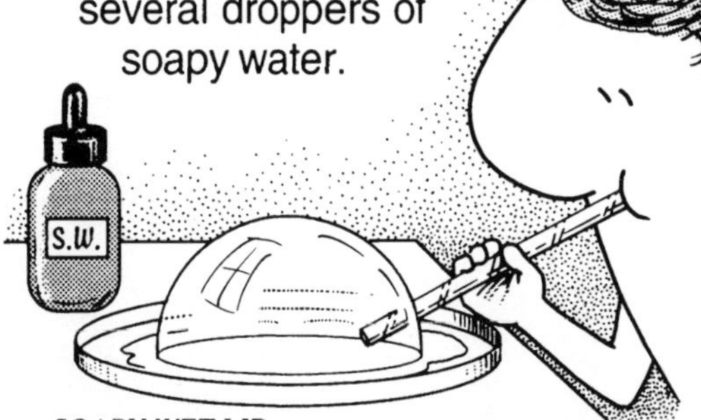

a. Dip your straw to the bottom of the soapy water bottle, then blow a dome-shaped bubble on the wet surface of the lid. (Pull out the straw before your dome pops.)

b. Describe how the surface of your bubble changes in reflected light. Can you predict when the dome will break?

INTERFERENCE

Cohesion / Adhesion (   )

1. Cut out the top rectangle from the Interference sheet. Graph how the *first* set of waves (upper + lower) add together to total one taller wave.

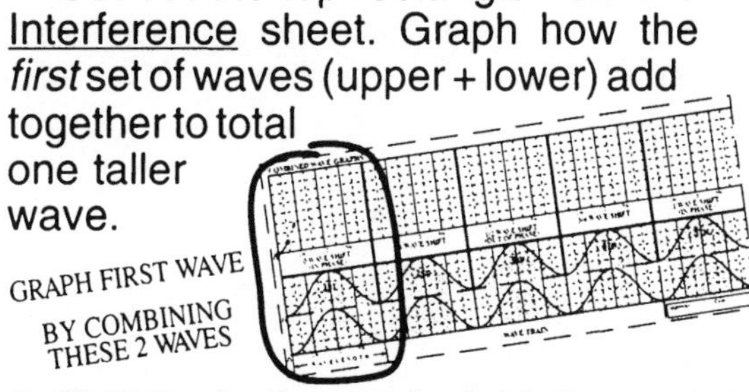

2. Cut off the bottom Wave Train on the dashed line. Shift it 2 squares *right* (1/4 wave length) and tape lightly. Graph how the *second* set of waves now combine.

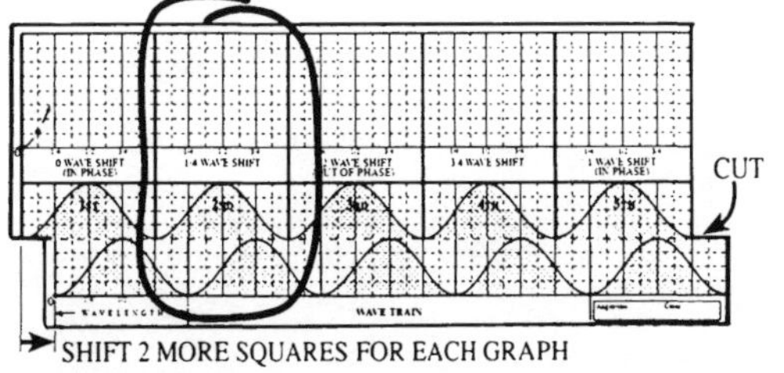

3. Shift the bottom train right, 2 squares at a time, to finish the last three graphs. (Save the bottom train for the next activity.)

4. Rephrase each statement below in terms of the graphs that you have drawn:
   a. Waves moving in phase interfere constructively.
   b. Waves moving out of phase interfere destructively.

5. *Coherent* (in phase) wave trains of blue light are shifted by these wavelengths. Complete the table.

WAVE SHIFT	0	1/4	1/2	3/4	$1\frac{1}{2}$	2	$3\frac{3}{4}$
BRIGHTNESS	*high*	*med*	*zero*				

WAVE SHIFT  Cohesion / Adhesion (   )

1. Cut out the WaveTrain strip on the Thin Film sheet. This represents a light wave enlarged about 40,000 times, to the same scale as this Wavelength Ruler.⟶

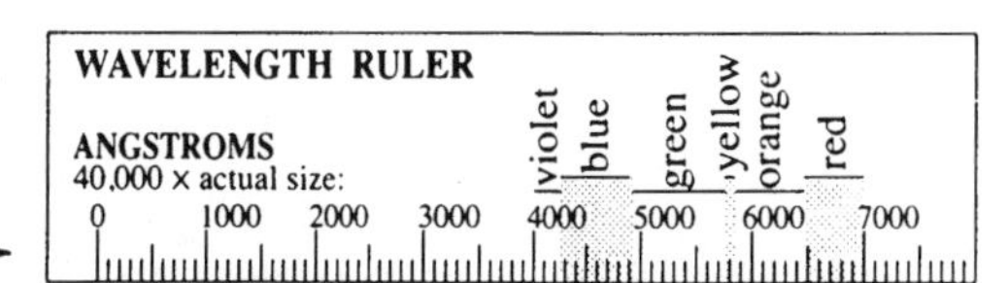

2. Measure the length of this light wave in Angstroms. Write this measurement, and the color we see at this wavelength, in the boxed space on the right.

3. Identify the wavelength and color of the wave train you graphed in the last activity. Fill in its box on the right.

4. Cut out the 2 Thin Film strips. Tape them end to end.

A B C D E F

THIN FILM
(40,000 ACTUAL SIZE)

   a. This shows a cross section of thin film. Why does it look so thick?

   b. Six rays of white light strike this film. Tell what happens to ray "A."

5. Use both wave trains like rulers. Which rays of white light...

   a. Interfere destructively to cancel blue? Yellow?

   b. Interfere constructively to reinforce blue? Yellow?

Be sure to measure both down and up.

COLOR BANDS  Cohesion / Adhesion 

1. Write your name on the lid of a small tub. Pour in bubble solution to cover the bottom.

2. Hold a paper cup upside down, and dip its mouth in the bubble solution.

3. Tilt the cup up at an angle so that light from a window reflects in the soap film. Observe what happens in the film.

   a. Describe in words and drawings how the film's surface changes over time.

   b. As gravity pulls solution downward, bands of color appear. Explain why this happens.

4. Repeat the experiment. When the top color band has drained a third of the way down the film, hold the cup level.

   a. Watch reflected light while blowing gently on the film with a straw.

   b. Describe what you see.

5. Cover your liquid and save it for all remaining activities.

LOOP THE LOOP

Cohesion / Adhesion (   )

1. Tie a snug loop of thread around a battery. Trim one end short; leave the other end long, then slide it off.

2. Pin your loop into an <u>Area Grid</u> backed by cardboard to form each figure below.

a. Count the squares in each regular polygon.

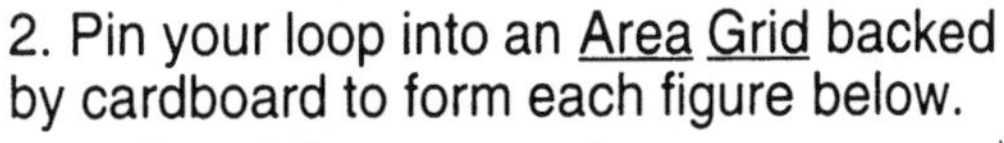

b. Which shape encloses the maximum area?

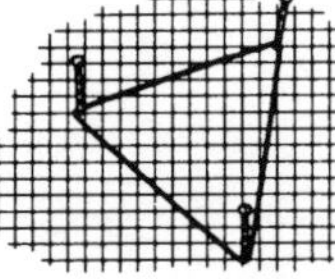

Shape	Area
2-PIN LINE ⁄	*(no area)*
EQUILATERAL TRIANGLE △	
SQUARE ▢	
REGULAR PENTAGON ⬠	
PERFECT CIRCLE ◯ *(trace base of battery)*	

3. Cover the mouth of a cup with bubble solution, as before. Set the cup upright on your table.

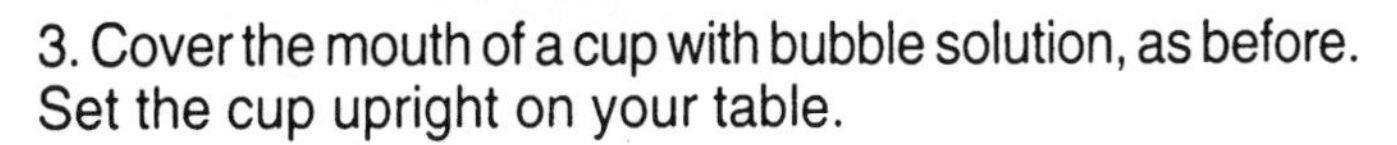

4. Wet your thread loop in the solution. Gently lay it in the middle of the film surface.

5. Break the film *inside* the loop with dry paper or a dry finger. What happens?

6. Explain in terms of ***cohesion***...
   a. ...what the loop's shape tells about the area of film that remains.
   b. ...why the film moves.

 21

MINIMUM SURFACES

Cohesion / Adhesion (   )

1. Dip the top of 2 batteries in bubble solution. Blow small bubbles of equal size on the top each battery through a straw.

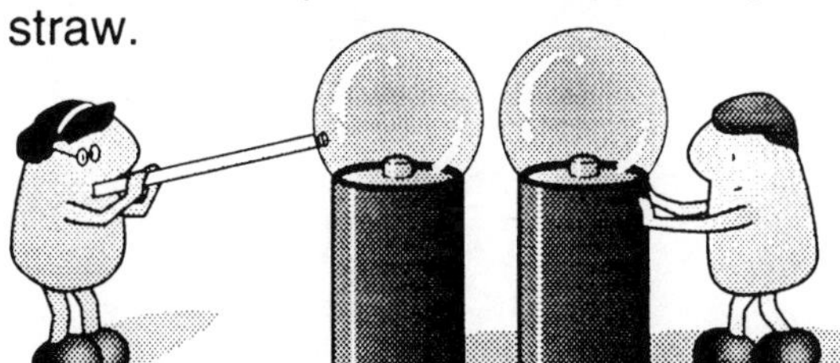

a. Why are these bubbles shaped like spheres, and not cubes?
b. Touch the bubbles together. Draw a side view showing how these films combine to reduce surface area.
c. Do bubbles of *unequal* size combine to form a *flat* common wall? Draw your findings.

2. Cut the bottom off a tapered cup. Wet it inside with bubble solution.

a. Dip the wide end in the solution to form a film over its mouth.
b. What happens? Why?

 22

BUBBLE ARCHITECTURE ○ Cohesion / Adhesion (   )

1. Carefully cut an even ring from a foam or paper cup about 3 cm ($1^1/_4$ inch) from the rim.

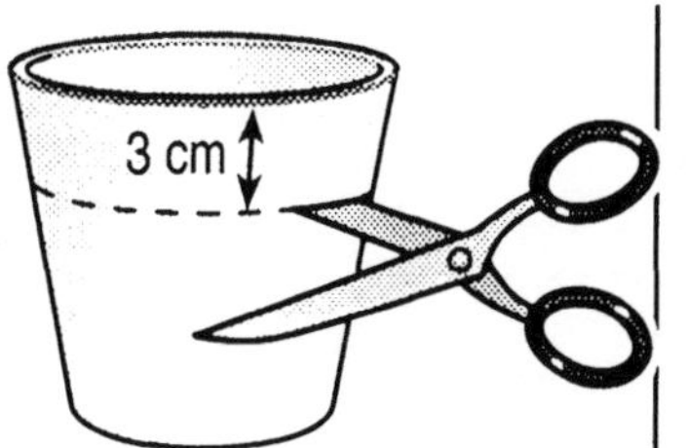

SMOOTH CUT EDGE

a. Dip the smaller end of this ring and the rim of a second cup in bubble solution. Touch the two film windows together, then pull them a little apart.

b. Draw what you see.

c. Break the center circle of film with a dry finger. Draw the film shape that results.

2. Use a straw to blow a layer of small connected bubbles on the wet lid of your tub.

a. Find these shapes between the bubbles: **triangle**; **square**; **pentagon**; **hexagon**. Describe what causes each shape.

b. State a general rule for making a regular polygon with **n** sides.

23

GIANT DOMES  Cohesion / Adhesion (   )

1. Cut aluminum foil and paper towel pieces the size of a 4x6 inch index card.

2. Fold each in half the long way. Join with masking tape, front and back, with the folded edges to the outside.

3. Roll the center tape around a narrow test tube. Fix it with more tape, leaving both ends free.

4. Slide your cylinder off the test tube, and soak the paper end in bubble solution.

5. Pour a puddle of bubble solution in the middle of your table. Blow into this puddle through the foil to make *giant* domes.

TABLE TOP

6. Measure the diameter of the ring left by the largest popped dome. Go for a record!

7. Can you blow giant domes through a straw? If you wrap a strip of soapy towel around the end, does this make a difference?

8. What variables influence dome size? Write a report.

24

WATER MOLECULES (ACTIVITY 5)

RIGHT

Twist shaded part DOWN

LEFT

Twist shaded part UP

A A B B C C D D E E

AREA GRID (ACTIVITY 21)

INTERFERENCE (ACTIVITY 18)

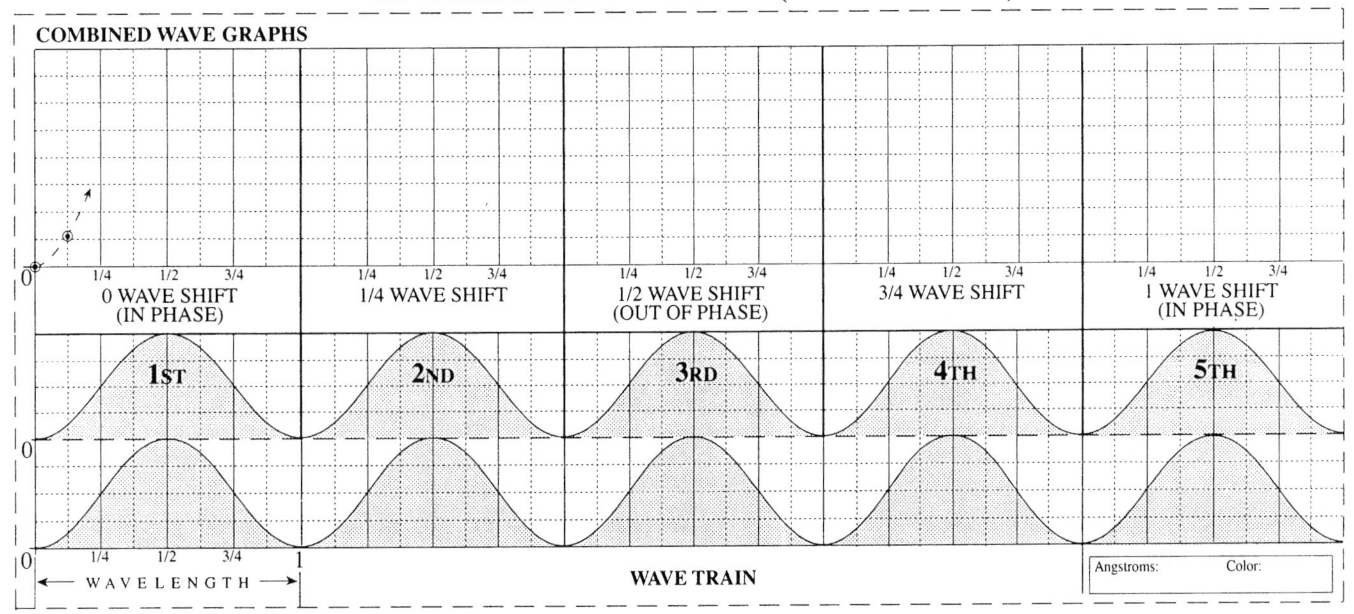

THIN FILM (ACTIVITY 19)

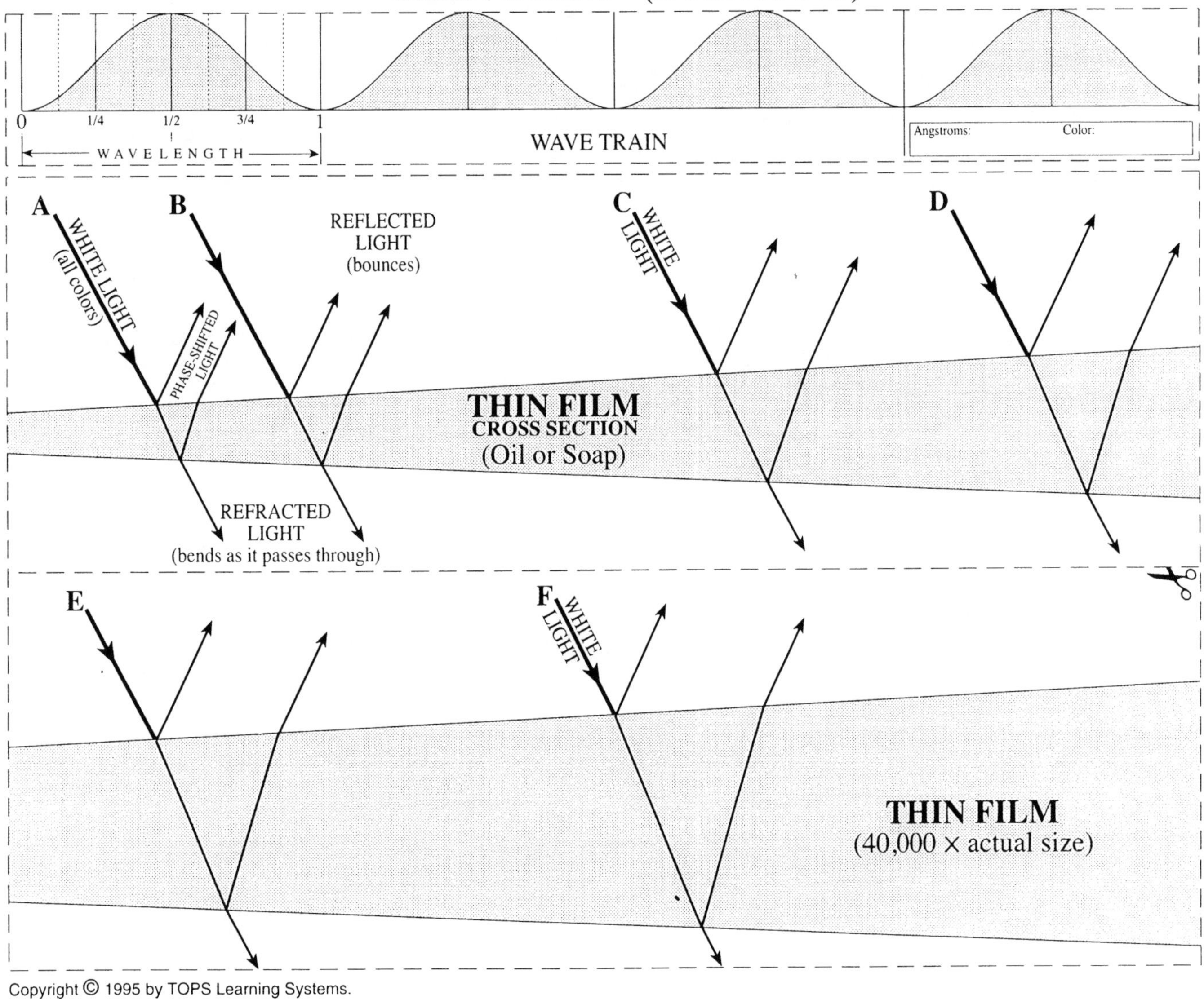

Feedback

If you enjoyed teaching TOPS please tell us so. Your praise motivates us to work hard. If you found an error or can suggest ways to improve this module, we need to hear about that too. Your criticism will help us improve our next new edition. Would you like information about our other publications? Ask us to send you our latest catalog free of charge.

For whatever reason, we'd love to hear from you. We include this self-mailer for your convenience.

Ron and Peg Marson

author and illustrator

Your Message Here:

Module Title ______________________ Date ____________

Name ______________________ School ____________

Address ______________________

City ______________________ State ______ Zip ____________

FIRST FOLD

SECOND FOLD

RETURN ADDRESS

____

____

____

PLACE
STAMP
HERE

TOPS Learning Systems
342 S Plumas St
Willows, CA 95988

TAPE HERE

Made in the USA
Monee, IL
03 March 2023